VAJONT

History for a Sustainable Future

Michael Egan, series editor

Derek Wall, *The Commons in History: Culture, Conflict, and Ecology*

Frank Uekötter, *The Greenest Nation? A New History of German Environmentalism*

Brett Bennett, *Plantations and Protected Areas: A Global History of Forest Management*

Diana K. Davis, *The Arid Lands: History, Power, Knowledge*

Dolly Jørgensen, *Longing and Belonging: Recovering Lost Species in the Modern Age*

François Jarrige and Thomas Le Roux, *The Contamination of the Earth: A History of Pollutions in the Industrial Age*

Tirthankar Roy, *Monsoon Economies: India's History in a Changing Climate*

Elena Kochetkova, *The Green Power of Socialism: Wood, Forest, and the Making of Soviet Industrially Embedded Ecology*

Marco Armiero, *Vajont: The Political Ecology of an Unnatural Disaster*

VAJONT

THE POLITICAL ECOLOGY OF AN UNNATURAL DISASTER

MARCO ARMIERO

The MIT Press
Cambridge, Massachusetts
London, England

The MIT Press
Massachusetts Institute of Technology
77 Massachusetts Avenue
Cambridge, MA 02139
mitpress.mit.edu

The open access edition of this book was made possible by generous funding and support from MIT Press Direct to Open.

CENTRO
PER IL LIBRO
E LA LETTURA

This book was translated with support from the Italian Ministry of Culture's Centro per il Libro e la Lettura.

The MIT Press would like to thank the anonymous peer reviewers who provided comments on drafts of this book. The generous work of academic experts is essential for establishing the authority and quality of our publications. We acknowledge with gratitude the contributions of these otherwise uncredited readers.

This book was set in Bembo Book MTPro by Westchester Publishing Services. Printed and bound in the United States of America.

Library of Congress Cataloging-in-Publication Data is available.
ISBN: 978-0-262-05301-3

EU Authorised Representative: Easy Access System Europe, Mustamäe tee 50, 10621 Tallinn, Estonia | Email: gpsr.requests@easproject.com

Dedicated to Tina Merlin,
who taught us to disrupt public order

CONTENTS

ACKNOWLEDGMENTS

Writing the short thank you notes for my work has been an enormous pleasure, time and time again. I find it wonderful to remember the many people who have supported me, in different ways, during this work of research and writing. In my work, I have often—unusually often for a historian—chosen teamwork, writing together with colleagues and friends. I remain deeply convinced that even when only one name is on the cover, books are always the result of many encounters, readings, and dialogues. The names of the many friends who have accompanied me in this encounter with *Vajont* include: Valeria Rucco, library manager at the Geography Institute of the Federico II University of Naples, who was always patient and knowledgeable; Michelangelo De Donà and Silvia Miscellaneo, who helped me access the rich archives of the Vajont trial; Toni Sirena and Adriana Lotto, who allowed me to get to know Tina Merlin a bit better; Elisabetta Novello, who generously filled the gap between me and the archives in Italy by helping me find the Lorenzo Rizzato trial papers; the Canestrini family, who guided me on the trail of Sandro's archive, and the staff at the G. Tartarotti Civic Library of Rovereto, who facilitated the possibility of accessing those papers; Giacomo Braulin, who guided me in my exploration of the Vajont Museum, part of his excellent master's research in environmental humanities at Roma Tre University; Enrico

Bacchetti, Chiara Bergamini, and the library of the Belluno Historical Institute of the Resistance and the Contemporary Age, who sent me an otherwise unavailable article; Tiziano Dal Farra, who is an invariable guarantee for unearthing materials on Vajont—his generosity toward me has remained consistent over the years; Enrico Biagioli, Giancarlo Balzarini, and Angelo Caponnetto, who took me along with them when, as very young scouts, they arrived in Longarone in the days immediately after the disaster; and Pierfrancesco Rizzato, who helped me get to know his father Lorenzo better, to whom I direct my deepest admiration. I also thank friends who listened and supported (perhaps even endured) me: Maria Cristina and Franco, Egidio and Eleonora, Agustì, Salvo and Federica, Gregg, Serenella, Mario and Patrizia, and Daniele and Ilenia.

I am fortunate to share my daily life with Stefania Barca, a scholar who is rich in new ideas and always revolutionizes one's perspective. I have no doubt that her presence in this book is much stronger than I realize. Giulia has been listening to my Vajont stories for years. Like me, she has a complicated relationship with Italy, but I am happy that she got to know Tina Merlin, Lorenzo Rizzato, Sandro Canestrini, and the disobedient Italy we are both proud of. I would also like to thank our nonhuman family member, Hope. He doesn't know it, but his company and care have done me a lot of good.

A final thanks to Andrea Bosco, who can envision a book in a phone chat and believe in it to the end, and to Beth Clevenger, who saw in this story the potential to reach a broader international audience.

Figure 0.1
Map of Italy/Vajont produced by Chiara Braucher.

PROLOGUE: THE FACTS

On October 9, 1963, at 10:39 p.m., a landslide of nearly three hundred million cubic meters of rock plummeted from Monte Toc in Northern Italy into the Vajont hydroelectric reservoir below. The impact produced an enormous wave of fifty million cubic meters of water that partly bypassed the dam and, rushing at a speed of a hundred kilometers an hour, brought death and destruction to the entire valley. The town of Longarone was almost destroyed. According to experts, the landslide released twice as much energy as that produced by the Hiroshima atomic bomb. There were about two thousand victims. The hydroelectric plant had been built by the Società Adriatica di Elettricità (Adriatic Electricity Company; SADE)[1] and had recently come under the control of the fledgling Ente Nazionale per l'Energia Elettrica (National Agency for Electric Power; ENEL) as a result of the 1962 law on the nationalization of electricity.

In 2008, the United Nations Educational, Scientific and Cultural Organization (UNESCO) included Vajont among five exemplary stories that can guide us toward a sustainable future. Indeed, Vajont was a disaster foretold, a history full of portents and predictions that ought to have prompted precautionary measures. In 2023, UNESCO reaffirmed the importance of the Vajont disaster as an exemplary story of poor risk management by including the Vajont trial archive in its Memory of the World program, dedicated to the promotion and preservation of documentary sources.

I will address all of these facets in the following pages. Here, at the beginning of our journey, I include a concise and undoubtedly incomplete chronology that I hope will help the reader navigate the Vajont swirl.

1926: The first rough draft of the Vajont Dam is produced by engineer Carlo Semenza.

1929: An application is submitted for the construction of a dam, with a maximum reservoir elevation of 656 meters above sea level.

1937: A new application is submitted, with a reservoir elevation at 660 meters above sea level.

1943: The Higher Council of Public Works approves the dam project in the absence of a quorum of voters.

1957: The project is raised to a reservoir elevation of 722.5 meters above sea level. The first report is produced by geotechnical scientist Professor Leopold Müller on the instability of Monte Toc.

1959: The dam is now completed. Geologist Edoardo Semenza, son of engineer Carlo, discovers the Monte Toc ancient landslide. In March, the Pontesei landslide occurs, just a few kilometers from Vajont. In May, journalist Tina Merlin publishes an article against SADE in the newspaper *L'Unità*, for which she will end up on trial and will later be acquitted.

1960: A landslide of seven hundred thousand cubic meters falls into the reservoir.

1961–1962: SADE commissions the Ettore Scimemi Center for Hydraulic Modeling (Centro Modelli Idraulici) to conduct hydraulic tests to examine the effects of a landslide in the reservoir.

1963: Control of the plant passes from SADE to ENEL. Despite the recommendation of the Center for Hydraulic Modeling, it exceeds seven hundred meters.

1963: The disaster occurs.

INTRODUCTION

October 9, 1997

It is, of course, not easy to remember where you were twenty-six years ago, especially if, all things considered, it was just another evening. We had just moved into a very small apartment in the Vomero neighborhood, leaving a beautiful old house in the center of Naples. One phase of my life, my doctorate in economic history, had ended. From that small apartment perched on the hill of San Martino, it was my turn to contemplate my life. Perhaps because of the dollhouse-like proportions of the house, I was afraid I had ended up trapped between the temptation of the discipline of economic history, still able to offer some job opportunities, and the desire to break free from the cage and study what, at that time, seemed to interest only me and no one else in the Italian academy.

Our television and my academic ambitions both seemed decidedly outsized for that new phase of my postdoctoral life, well represented by the new apartment—small but "normal" compared with the previous one that had lent itself well to nurturing all kinds of dreams. On that evening of October 9, 1997, we were there glued to the screen, tuned to channel 2 on Italian public television, which was broadcasting—amazingly, in prime time—"Vajont 9 Ottobre 1963," a two-hour monologue written by Marco Paolini, who was also on stage, and Gabriele

Vacis. Thirty-four years had passed since the Vajont tragedy, and public television decided—finally—to honor the victims of that disaster with a civic oration that transformed the dam into a theater of memory. We were not, in fact, alone: Paolini's show subverted all predictions by garnering some 3.5 million viewers. Despite being a monologue and, moreover, recited by an actor who was not even particularly famous to the mainstream public, the Vajont oration captured the attention of many by demonstrating something simple and beautiful: Storytelling is a powerful act, especially when the story being told sabotages mainstream narratives and liberates suppressed memories.

At that time, after a complicated PhD in economic history that had certainly not pushed me toward any disciplinary transgression, I had not yet read either Michel-Rolph Trouillot's *Silencing the Past* or bell hooks's *Belonging: A Culture of Place*.[1] My doctoral thesis had been criticized for being written like a novel—apparently, writing badly was an unmistakable sign of scholarly rigor—and I was continually reminded that dealing with the environment was a laughable thing, perhaps good for an activist but not for a serious and impartial academic. During that fall of 1997, exactly one year after earning my PhD, I needed to rediscover the reasons to be a historian and, in particular, to do it on my own terms, to continue to deal with the environment, without losing my political commitment and my desire to talk with all kinds of people. Glued to the couch on that October evening twenty-six years ago, I just remember thinking that it was necessary to tell stories such as the Vajont story and to do it the way Paolini did, siding with those who had had no part in the national memory, who had been left out of grand narratives about the economic miracle and the country's modernization. The other feeling I distinctly remember is a sense of shame at knowing so little about that history. Even I, a historian by trade—precarious, of course, but still working with the past—knew little about Vajont. How could it be that in five years of high school, and just as many years of a humanities degree—in my case with a focus on contemporary history—there had been no trace of Vajont? Indeed, it had taken a theater actor to restore Vajont to our collective memory.

My career as an environmental historian has been marked by events such as this. I had become fascinated with forests after hearing the song "Geordie" in the version by the Italian folksinger Fabrizio De André, chasing down research trails about property and nature, customary uses and the law, trees, farmers, and foresters.[2] Paolini's monologue did not take me straight to Vajont—I would not write about this disaster until many years later in my book *A Rugged Nation*[3]—but it reassured me: Not only was there a need to tell the intricate stories that link humans and the environment, but there was also the possibility of using memory as a barricade to defend against the flood of toxic narratives that invisibilize and normalize injustice.

The next day, I did not immediately take up working on Vajont, but Paolini's civic oration, like De André's song, had a profound effect on my way of doing environmental history.

Be it a dam in the Italy of the economic miracle,[4] a forest poised between privatization and the commons, or a landfill on the edge of the contemporary metropolis, the environmental history I wanted to practice would always be concerned with power, inequality, and conflict. And I would continue to nurture the ambition to speak to many people. Fortunately, my doctorate program was over, and I could return to writing as I wished, without the limitation that academic rigor should be inversely proportional to the accessibility of the writing.

SCIENCE WILL SAVE US

In *A Rugged Nation*, I described Vajont as an exemplary story for understanding the relationship between the environment and Italian society, at least in contemporary times. One cannot understand Vajont without connecting the geology of Monte Toc, the economy, and SADE's political power; or the marginality of the Italian mountain communities and the subalternity of the Italian academy to economic and political power; or the hydraulics of a large reservoir and the mechanics of state control systems over major public works. To those who imagine environmental history as a neutral field of study, dedicated to nature and thus free from the partisan divisions of the social, a story such as Vajont reminds us that ecology is always political—even, perhaps especially, when its stories are naturalized, as if the deaths of two thousand people were merely the tragic result of an ancient landslide. Perhaps it may sound counterintuitive, but my version of environmental history does not rhyme with naturalization. I never believed that the challenge for environmental historians was to explain everything through nature—geographic determinism is enough for that. Rather than distilling the components of environment from the complicated blend of socioecological relationships, I preferred to try to understand the mixture. In short, in my version, environmental history is not about separating the environmental from the social. Rather, it is about seeing the socioecological relations our world is made of. A dam is water becoming energy, capital embodied within the landscape, the ghosts of those left metaphorically or materially submerged, technology and science deciding the life and death of thousands of people and entire ecosystems.

UNESCO expressed similar concerns in 2008 when, on the occasion of the International Year of Planet Earth, it decided to include Vajont as one of five exemplary stories to keep in mind for building a sustainable future. The UNESCO document spoke of the Vajont disaster as an example of the failure of interdisciplinary cooperation between engineers and geologists in designing major infrastructure. In the chapters that follow, we will see the details of this strange failure, with the dam still standing and the mountain crumbling. UNESCO used Vajont to emphasize—and rightly so—the need to integrate different expertise when undertaking major infrastructure projects. Undoubtedly, one cannot come to understand complex ecological systems by remaining imprisoned in a rigid division of knowledge that invariably fails to understand the system and how it works. Vajont, indeed, seems to be a textbook example of this inability to combine multiple kinds of knowledge, leading to the construction of an incredibly solid dam but at the foot of a "rotten mountain" (the meaning of "Monte Toc" in the local dialect). Although I agree with the multidisciplinary agenda proposed by UNESCO as a strategy to overcome disciplinary silos and address the socioecological crisis, I believe that one cannot understand Vajont simply as a technical scientific failure. Faced with the multiple socioecological crises of our times—from pandemics to climate change—we often evoke the authority of science and defer to the technical solutions that can flow from it. The debate seems to polarize between those who believe in science and deniers of all stripes, ready to mobilize against vaccines or doubt the most basic scientific notions. There is an almost Enlightenment idea of knowledge and a rather naive idea of power: Seeing the nature of the problem is fundamental, but this does not free one from the power relations that determine the possible responses. It is not enough to see that a factory pollutes to free a worker from job blackmail, and it is not enough to see the harms of fossil fuels to convince people living in the suburbs of metropolises not to use their cars. If there is one thing Vajont teaches, it is that science has a lot to do with power: the legitimization of knowledge, the relationship with political and economic power, the conflict between opposing sets of knowledge and interest, and the uncritical and unconscious adoption of pervasive ideologies such as modernization. These are central concerns within the Vajont disaster that I believe have universal value. For this reason, in the following pages, I will focus mainly on these points rather than on the technical and scientific failure of the hydroelectric project. To make the inquiry more explicit, my question is this: Was it a lack of multidisciplinary knowledge that killed two thousand people, or was it hydroelectric capitalism—the imperative to maximize profits for a few and minimize precautions for many? If, as we shall see in the next chapters, there was no lack of knowledge that might have alerted authorities and the corporation to the threat, Vajont confirms, in the tragic finality of a massacre, that it is not knowledge that will free us from the socioecological crisis.

Since I have no reputation to defend as a neutral scholar, I can freely say that "only the struggle pays" (*solo la lotta paga*). To elaborate further on this slogan, which works fine as it is, I would say that Vajont demonstrates, in a material fashion, not so much the need for collaboration between disciplines but rather the fact that science is, often, a battleground. Indeed, Vajont itself reminds us that things get even more complicated when it comes to deciding what we legitimize as knowledge and what, instead, becomes superstition, the irrational fears of ignorant hillbillies. Speaking to a reporter who asked her reason for not leaving Longarone, despite widespread fears about the dam, one survivor remarked, "Because we hoped they [the experts, the authorities] knew more than we did." Authority and legitimacy were all placed on one side, with the experts possessing the right academic degrees, the proper technical language, and modern tools tasked with monitoring the situation. During the COVID-19 pandemic, I was accused of expressing critical opinions about the singular way government officials in Sweden managed the emergency, characterized by very few restrictions, without holding a doctorate in epidemiology or medicine myself; in short, without being an expert. Vajont poses the question of the delicate balance between professionalized expertise, public or common knowledge, perceptions of risk, and public debate. I am glad the doctors in the intensive care unit at the Stockholm hospital to which I was admitted didn't convene an assembly deliberating whether to intubate me. I remain convinced, however, that, in some cases, when entire communities are affected by potentially risky projects, there is a need for what Silvio Funtowicz and Jerome Ravetz call "post-normal science," which calls for an expansion of expert opinion and a democratization of science.[5] This is not to legitimize anti-scientific attitudes but rather to recognize the plurality of situated knowledges, useful in achieving what feminist philosopher Sandra Harding has called "strong objectivity.[6] In the case of Vajont, as with so many occupational diseases, adopting a position of listening produces advances in knowledge. Sometimes, being biased is the best guarantee of being a good scholar.

THE EYE OF THE LAKE

Overlooking the Vajont Dam, it feels as though we are touching the much-debated Anthropocene, in spite of its official rejection.[7] The scientific definition insists on the planetary scale of the transformative action of humans. Yet, Vajont functions as a microcosm in which so many of the central questions of the Anthropocene are reflected. At Vajont, humans have become a geological force, capable of moving mountains and putting an ancient landslide back in motion. Engineers wander around at Vajont, powerful as gods, reckless as only gods and heroes can be. At Vajont, everyone seems to know that disaster is imminent, and yet it is never real enough to elicit immediate action. The Anthropocene as seen from the Vajont Dam is the story of impending disaster caused by human hubris, a story where omens and signs are never enough to halt the inexorable march of progress in the name of precaution.

Perhaps we might get a different version of the Anthropocene by looking at the events from the perspective not of the Vajont Dam but rather of the cemetery where the victims of the disaster are buried. Of course, the plot of the story would not change. The millions of tons of rock hurled from Monte Toc into the Vajont basin would cause the same wave and the same deaths. But looking at this story from the Vajont cemetery, we might be less convinced that it is human hubris and technology that are the culprits of the disaster. Viewing the Anthropocene from the Vajont cemetery, the responsibilities of the species seem narrow to us. On those tombstones, we don't find the names of the great executives of hydroelectric capitalism or even the civil servants who put themselves at its service. Those who pay the bill for the disaster are not some unfortunate people drawn by lot in the great lottery of history. Rather, they are subalterns, marginals, communities that are worth nothing except as sites of value extraction for the profit of a few, in the name of the purported general interest. In my book *Wasteocene*,[8] I wrote that, from the Vajont cemetery, as from so many other cemeteries in the sacrifice zones scattered across the peripheries of affluence, what we see, more than the Anthropocene, is the Wasteocene, an age dominated by socioecological relations that produce wasted communities through processes of value extraction and othering. If, unlike the Anthropocene, which is so profoundly planetary, the Wasteocene is instead embodied, localized, and experienced, then, indeed, the Vajont microcosm is a perfect reproduction of it to scale.

It is difficult to pass through the eye of a needle, especially if you are rich—Christians are well aware of this.[9] What passed through the eye of the Vajont "lake"—and perhaps a lake (*lago*) really does resemble the eye of a needle (*ago*)—were not only two thousand victims but also the history of an entire nation. Some might claim a miracle: The camel made it through—Italy became modern thanks to hydroelectricity. If someone was sacrificed, better to forget it so as not to spoil the celebrations. In the Vajont parable, however, there is no salvation. The poor—that is, those on the margins of national history—remain there. For them, the eye of the lake is not the golden gateway to some paradise but rather the siphon of a gigantic basin that sucks them into a whirlpool of mud and water from which not even memory seems to be able to snatch them. For if it is true that the eye of Vajont sees the history of Italy pass through, it is also true that this history has remained substantially hidden. At Vajont, the Gospel is twisted. Passing through the eye of the lake, the rich are saved and the poor end up condemned to hell.

GLIMPSES OF THE FLOOD

In this book, I will try to narrate Vajont as an observatory from which to look at the history of Italy and the Anthropocene. I will do so by following the gazes of the many who have traveled through the history of the dam, with the awareness that those gazes have never all had the same legitimacy and authority in each case. We will see Vajont through the eyes of the mountain people who listened to the voice of Monte Toc, awaiting the disaster. We will see the Vajont valley as a perfect hydroelectric engine through the eyes of experts such as geologist Georgio Dal Piaz and engineer Carlo Semenza. Tina Merlin will lend us her gaze to see Vajont through the eyes of a communist journalist capable of seeing the great contradictions of industrial capitalism within that tiny piece of the world as well as its entanglements with political power. We will try on the thick glasses of Mario Fabbri, the prosecutor in the Vajont trial, who brought the top executives of the Adriatic Electricity Company to the dock with the responsible government bodies. Finally, we will ask what remains of Vajont in collective memory—or to retain the metaphor of the gaze, we will seek to understand how the nation looked at Vajont by attending to how its memory and forgetting have been organized.

1

THE FLOOD WITHOUT AN ARK

October 9, 1963, 22:39

Many people have experienced the unsettling sensation of being immobilized in bed, often upon just waking. Experts call this phenomenon “hypnagogic paralysis,” but to most, it is simply a bad feeling that lasts a few moments at most. For Micaela Coletti, president of the Vajont Survivors Committee, however, it marked the onset of a catastrophe.

As her grandmother instructed her sister to close the shutters against what seemed like an approaching thunderstorm, Micaela was sucked into the chaos of the Vajont disaster. She recounted those harrowing moments in an interview gathered by the committee:

> A gust of wind from the distance, slamming the shutters. Then . . . a dull, deep noise. The sensation that the bed was picking up speed, a terrifying force gripping my back, bending me in two, crushing me. It felt as though my body had turned to rubber—stretching, shrinking. My eyes became like two stars; an immense pressure yanked me by the hair, pulling me into an endless pit, pinning my arms to my sides, leaving me completely immobile. A sharp pain radiated from the base of my back, and breathing seemed impossible.
>
> This overwhelming force held me tightly, but it made me angry! I remember thinking, *“No! I won’t let go!”* even though it felt inevitable. Somehow, I managed, with immense effort,

> to raise an arm and touch my face, searching for my eyes, nose, mouth. I felt thin, squashed, without any dimension. Slowly, I raised my arms above my head, reaching for something—anything—to hold onto. And then . . . blackness. Complete, total blackness.[1]

Despite feeling crushed and nearly immobilized in her bed, Micaela was dragged at least four hundred meters from her home. Similarly, Giuseppe Sacchet was discovered far from his house, still lying on his mattress amid the rubble. In his interview about that fateful night, Giuseppe transformed his mattress into a flying carpet, describing how it carried him over the raging waters of the Vajont reservoir.[2] Renzo Scagnet also found himself far from home, completely naked on the steps of the town hall. His sister was found dead thirty kilometers from home, his father's corpse in the village of Fortogna.[3] Matelda Coletti was swept into a giant sinkhole that had opened in the floor of her home. Dragged by the torrents of water and mud, she became trapped, her foot pinned beneath rubble, and suffered a head injury that partially paralyzed her face.[4]

The fortunate ones, who did not end up buried under the rubble, recount the horror of witnessing the disaster up close. Luigi Rivis, who was eighteen years old and already working as a surveyor for SADE, shared his memories during an interview with a local radio station. Luigi recalled how he first understood the magnitude of the event. Initially, he was reassured by the sight of only a few emergency vehicles leaving Longarone, believing the damage to be limited. However, when he ventured to the banks of the Piave River with flashlights, he came across "a body, naked, though not whole," caught in the bushes. Reflecting on that moment, he added, "That's when we realized something enormous had happened."[5]

Italo Filippin was also among the fortunate ones who managed to reach safety. As he recounts in a brief memoir about that night:

> I remember a chaotic escape, the screaming . . . Someone took the initiative to light fires, around which the survivors began to gather and check on each other. . . . When dawn came, it felt as though we were in a different place; the valley was unrecognizable. Many hamlets had been completely wiped out—nothing was left. The lake level, now filled by the mountain, had risen by nearly twenty meters, if I remember correctly. Everything was floating on the surface: corpses, rubble, household goods.[6]

For another survivor, Gino Bonsembiante, the memory of Vajont was also linked to the mutilated bodies—without arms and heads—etched in his mind. Like the monster Micaela had once imagined under her bed, Vajont had devoured its victims, spitting out their remains into the river of mud and debris it left in its wake. Maria Pia Bassetto, a young postal worker on duty in Longarone, recalls the rescuers emerging from an underpass that led from the Piave to Fortogna, carrying the mutilated corpses recovered from the riverbed on their shoulders.[7] All around were the bodies of animals that had died in the disaster, their bellies bloated and their legs in the air.[8] In his memoir about the Vajont disaster, Bepi Vazza describes the same horror—a macabre forest of death, human body parts hung from tree branches:

> I remember rummaging through the branches of a large plant with its roots in the air. We saw an arm sticking out from the branches and gravel. We made some space around it and tried to

> pull it by the hand, hoping that the rest of the body would come with it. Instead, it was just a woman's shoulder, detached from the rest of the body. We called some soldiers who were nearby to help us, but there was nothing left to do.[9]

It took only a few minutes to radically alter the landscape between the dam and Longarone. All the survivors describe the sense of alienation that overwhelmed them when they first looked out over the ruins of the disaster. In the documentary video produced by the *Corriere delle Alpi* newspaper to mark the fiftieth anniversary of the disaster, Bruno Pradella and Rita Marogno shared the memory of their bewilderment when, in the moments after the disaster, they found themselves in an unrecognizable environment. Nothing resembled the world they had known just minutes before.[10]

But the Vajont flood was not only experienced with the eyes. All the survivors remember the noise—an indescribable sound—that accompanied the landslide and the subsequent flood. Gervasia Mazzucco, who was eleven years old in 1963, provides the best description of the noise from that night: "It's not like you could hear the noise from the other side of the mountain; you're inside a huge thing, and my grandmother said, 'It's the end of the world.'"[11]

Gervasia's recollection is particularly powerful because, in just a few words, she manages to convey the kind of noise unleashed by the landslide: a noise that, although it comes from the mountain, doesn't stay outside but enters people, absorbs houses, and becomes like a colossal blob incorporating everything. The voice of Gervasia's grandmother announcing the apocalypse seems to merge with the indescribable noise of the flood, as if this elderly mountain woman had found the key to translate that unknown sound into human words.

In the memories of other survivors, however, the supernatural noise of the landslide blends with much more human sounds. Vittore Zannol, for example, associates the deep sound of the Toc plunging into the lake with the desperate cries of children from the house next to his own, calling out to their mother for help.[12]

Noise and wind moved together on the night of the Vajont disaster. Bepi Vazza recounts the "deadly, impossible wind" of that night—a wind that penetrated the mouth, the nostrils, the eyes. Like Gervasia, Bepi describes a disaster that breaks the barriers between the world outside and the human body; water, mud, noise, and wind shatter every boundary, enter everywhere, and incorporate everything. "The earth trembled and jerked out of proportion," Vazza writes. "The noise grew louder and louder; you couldn't hear it with your ears, it came up from the earth, entered your feet, and traveled through your body until your head felt like it would explode. It was almost impossible to bear."[13]

The wind was so strong that it took one's breath away, as Elsa Capraro, who turned twenty-two on October 9, recounts. Elsa believed it was fear that took her breath away as she recited a prayer under the arch of the door, preparing for death. Only later did she realize that it wasn't fear that was responsible for her apnea but rather the air displacement caused by the water being hurled at full speed into the Vajont canyon.[14] Strong winds blocked Bepi Vazza as he tried to reach his home and rescue his mother. He felt his clothes being ripped off his body, and then he was lifted up and hurled away. It was another survivor who rescued him, preventing the wind from dragging him farther away.[15]

After so much noise, a surreal silence greeted the first rescuers to arrive at the Vajont ruins.

Figure 1.1
Aerial view of Longarone, before and after the disaster. Courtesy of the Vajont Survivors Committee.

RESCUERS

At 1:00 a.m., roughly three hours after the disaster, Aldo De Col, a firefighter on duty in Longarone, sat on the ruins of his family home and noticed that, all around him, there was a "tomb-like silence" that not even the strong wind could break.[16] Giancarlo Zanarini, who had driven from the Faesite (hardboard) factory not far from Longarone to the disaster site, also recalled: "Climbing through rubble and lumber, I arrived at the bridge over the Maè River, but the road no longer existed. Only the sound of the Maè's water could be heard, a grave silence and darkness filling everything."[17]

The first rescuers were survivors—locals who had escaped the fury of the waters—wading through the rubble, trying to detect the faintest signs of life in order to get others to safety. Everyone remembers the surreal silence, the deep darkness, and the feeling of radical strangeness in a landscape that was supposed to be familiar. Giancarlo Zanarini, for example, struggled to get his bearings, unable to recognize anything that night, until a car along the road pointed its headlights at the valley, illuminating the Pirago bell tower—the only building that still seemed to be standing. Zanarini was only slightly mistaken: A 1964 fire department report added a house, which had been saved because it was built on a hill, as well as a gutted little church.[18]

Giancarlo Zanarini's journey through the night of the Vajont disaster takes on Dantean overtones, becoming almost a metaphor for a collective experience—a journey to hell without redemption or guidance. Zanarini witnessed the flooding from the vantage of the Faesite factory, which was spared from the shock wave thanks to the protection of a forest that dampened the violence of the water. The actual Dantesque journey began right there in the factory when Zanarini found parts of a woman's corpse on the laboratory staircase. That grisly discovery, which occurred in the deep darkness into which the factory had been plunged, opened the door to the real Vajont inferno. Zanarini realized the disaster must have been massive—there were no women among the factory workers, meaning the body must have come from outside—and, along with four colleagues, he set out for Longarone.

At the Faè bend, heading toward Longarone, the contours of the disaster began to take shape. The workers found the road cluttered with rubble, and among the debris, the first corpses appeared. The sight of a sixteen-year-old girl standing in the middle of the road added an almost surreal touch to the scene. But the screams of the injured coming from beneath the rubble quickly brought them back to reality, and the workers began digging with their hands. Yet, Zanarini did not stop but rather continued his journey, eventually arriving, as mentioned earlier, in Pirago. And it is here that the gates of hell opened. Zanarini recounts:

> I climbed up to the bell tower, the only building left standing. I didn't know there was a cemetery nearby; when I crossed it, I

thought I was having nightmares, because the recent dead were mixed with the old. The water had torn apart dovecotes and niches, the crates were smashed, and skeletons were scattered everywhere, along with the remains of people swept away by the wave. I had a moment of almost panic, not understanding if I was dreaming, if it was the apocalypse, and if I was dead.[19]

One can hardly imagine the scene described by Zanarini. The familiar landscape of a small hamlet in Longarone was unrecognizable, so distorted by the waters that Zanarini did not even realize he had ended up in what was, until a few hours earlier, the cemetery of Pirago. In this moment, our guide through the Vajont night is correct: Surrounded by skeletons and human remains, Zanarini realized he was no longer in Pirago but had entered another world—perhaps one of those paintings in which artists of earlier centuries imagined the end of time.

Zanarini's journey through the apocalypse carried him among the dead, old and new; yet his encounters with the living were no less harrowing. Like any self-respecting guide, Zanarini felt the weight of what he had witnessed and the difficulty of sharing it with the living. On the road from Pirago to Longarone, he met Aldo De Col, the firefighter who had been listening to the silence on top of the ruins of his house. Zanarini told him that the entire village was gone, only to discover, when De Col knelt to weep, that he had just informed him of the death of all his family members.

Later, as Zanarini continued his wanderings through the night of the apocalypse, he made a different choice:

At the escarpment on the Maè bridge, I heard the voices of Mr. Dalla Mora, the deputy director of Faesite, Mr. Austoni, and Mr. Pellizzari, an employee, talking as they tried to get back up. I hid because I did not have the courage to say what I had seen [the

> director's villa in Pirago was completely destroyed]. I let them pass; I knew the director had a wife and children in.[20]

Hiding in order to avoid telling what he had seen, Zanarini did not want to be the one to bring the factory manager the tragic news of the end of the world. After all, it was the end of the world—of their world—that occurred that night at Vajont. It seemed better to hide than to have to find the words to describe the nothingness he had seen, or to have to tell the truth in a night that left no room for lies. Zanarini did not remain hidden for long. He soon resumed his descent into the underworld to encounter yet more pain.

Among the rubble, he rescued a badly wounded woman, who immediately tried to free a little girl—perhaps a sister—but the girl died in Zanarini's arms. Like any genuine journey to the underworld, this one too must end with a "then we emerged to see the stars once more" (*a riveder le stelle*) of some kind.[21] For Zanarini, it was the much more prosaic light of a flashlight. On the road to Faè, by now at 3:00 a.m., the light of a flashlight illuminated his face. It was General Ciglieri with two hundred Alpine soldiers—the first to be rushed to the scene.

"Longarone no longer exists," said the worker, but for him, having lived through it, it was not easy to speak of the apocalypse—or perhaps it was not easy to hear it told. Zanarini recalls that Ciglieri first thought he was dealing with a "disturbed" person, perhaps in shock. Only when he identified himself as a former Cadore Division officer did Ciglieri believe him. The apocalypse is not easy to recount, even for those who lived through it, and certainly even harder to believe for those who hear the tale.

The Alpini of the Cadore Brigade arrived at Vajont around 3:30 a.m., among the first rescuers to reach the disaster site—perhaps more accurately described as the crime scene—preceded only by survivors who became rescuers and some units of the Finance Police (Guardia di Finanza). In the days that followed, the military presence at Vajont grew exponentially to include about five thousand men, all under the command of General Ciglieri.

The journey of the Alpini soldiers to Longarone had been no less troubled than Zanarini's night pilgrimage. Italico Cauteruccio, a company captain in the Cadore Brigade in 1963, recounts this other part of the Vajont night:

> On the evening of October 9, the Regiment was put on alert as something serious had happened at the Vajont Dam. . . . The column had to stop at Polpet, near Ponte nelle Alpi, as the roadway along the right bank of the Piave River was impassable for vehicles. So the Alpini had to continue on foot, through the deepest darkness, while terrified locals spoke of an immense wave of the river that, with deafening noise, had overflowed its course. . . . It was beginning to dawn, and as we proceeded through an environment that seemed increasingly unreal, we could glimpse, here and there, the swollen silhouette of some dead animals and shapes that, at first glance, appeared to be piles of mud. But they were decomposed, completely denuded human bodies, which the wave of the river, with its immense force, had swept over, killed, and abandoned at the edge of its course. . . . Finally, we came to a deserted and ghostly heath: Longarone had disappeared, turned into an immense riverbed, with very few houses left at the edges, visibly half-destroyed or damaged. The first image that aroused extreme wonder and deeply struck us was, on the right, the intact dam, gray, severe, and imposing as always, embedded in the sides of the mountain—those, yes, flayed and eroded by the immense, impetuous wave that had swept over them.[22]

Bodies mixed with mud, unrecognizable, shreds of human bodies becoming debris, gruesome fruits hanging from trees—always naked, testifying that the disaster had dehumanized them, depriving them of their dignity. These frightening corpses are the protagonists of all the tales of that Vajont night. After a flood, as the waters recede, the stumps—the severed logs—often reemerge. Firefighter Giovanni Boite Stella realized that, at Vajont, those stumps were human bodies: "mothers hugging their children, pieces of people, corpses caught in the branches of the beech trees."[23]

The testimonies collected for the fiftieth anniversary of the disaster by the Vajont Foundation all return to this dramatic landscape of death, made up of bodies and human remains. Firefighter Luciano Giurato declared without hesitation that he had never seen so many corpses.[24] For Alpine soldiers Luigi Fontana and Giovanni Lanzarotto, the memory of the Vajont disaster was primarily marked by the vast number of bodies "scattered everywhere," forming, under the pylons of a bridge over the Piave River, an almost posthuman dam "of trees, beams, and furniture, even dead livestock and human corpses."[25]

Once again, it was corpses—of both humans and animals—that greeted the technicians from the Veneto telephone company, who had set out from the town of Conegliano for Longarone. Before their eyes flowed a ferocious Piave, laden not only with water but also with bodies, debris, cars, and even high-tension pylons.[26]

Here was the flood, not the universal flood but the particular one—the end of a world, with its memories and its belongings. Dino Milan, a member of the military contingent that rushed to Vajont, is unequivocal: "These people saw the end of the world," he declared in his testimony collected for the fiftieth anniversary.[27]

Pylons and cows, humans and cars, all mingled in the great blender of Vajont, and before the eyes of the workers, the Piave River carried disjointed fragments of modernity and tradition—like the frames of a film made of flesh, rock, memories, and kilowatts. It is no coincidence that, alongside the stripped and mutilated bodies of the victims, the other recurring image in the memories of Vajont is that of crumpled rails, which frequently appears in the accounts of witnesses. Angelo Caponnetto, one of the scouts who came to Vajont from Umbria, a hilly region in central Italy, recalls the amazement he and his comrade Giancarlo Balzarini felt upon seeing the rails at Longarone station, curled up like ribbons on a gift box.[28]

The insistence with which so many return to the tangled rails, I believe, not only reflects astonishment at the force of the waters—capable of bending the steel of the railways—but also conveys a vivid perception of the failure of modernity. Much more authoritative narratives might have attempted to impose a different view, one in which the dam, almost intact despite everything, would have symbolized the success of modernity. But for those who walked through Vajont that night, there was no room for such stories. Modernity was there, crumpled by a tidal wave, and the dam in the background seemed more like a threat than a promise of the future.

Not everything was visible through the Vajont night. Some images surfaced in the churning waters of the Piave, whereas others remained submerged—unseen but no less threatening. The modernity that arrived between the Vajont Gorge and the mountains of the Erto and Casso villages was also this: 104 barrels of isocyanate, or cyanide, that the wave had swept away from the Faesite factory and dumped into the macabre dance of the flood, along with everything else. The chemical disaster was narrowly averted, with 103 cyanide barrels recovered. Only one remains missing—yet another ghost left to haunt the Vajont night.[29]

Arriving in Longarone and the surrounding hamlets, the corpses that had crossed the path of the rescuers became an even stronger and more tragic presence. As soon as he arrived on the scene, Alpine soldier Sesto Fiocchi was taken by a priest to a church where eighty corpses awaited, ready to be loaded onto stretchers and transported elsewhere for identification.[30] The sheer number of bodies, as well as their condition, left an indelible mark on the memory of firefighter Ulderico Quintabà, who had rushed to Vajont from the Marche region. Quintabà recalls a 'pile' of corpses, stripped and mutilated..[31]

Younger rescuers also arrived at Vajont, including scouts, many of them little more than boys, thrown into that provincial apocalypse. Enrico Biagioli was sixteen years old when he arrived at Vajont two days after the disaster. He proudly tells me that it was their choice to serve at the cemetery; they could have been assigned to food distribution, but they decided they had not traveled a thousand kilometers "to go up to Longarone and make sandwiches."[32] Truckloads of bodies, often just human remains, arrived at the cemetery. "A truck arrived, they gave us a bag, we opened it, and found a man's foot—and still, I get chills because I dream about it at night—the head of a little girl, blonde, with her ponytail tie still in place, earrings, blonde hair, and blue eyes."[33]

The Vajont dead continue to haunt the dreams of those who encountered them, perhaps too young, like Enrico, or like a fellow scout who broke down in tears on a beach during a vacation because the image of a young girl's head floating in the water dragged him back into the whirlpool of Vajont, where the water showed no mercy and young girls could no longer play.[34]

Although these are extremely personal, almost intimate recollections, they essentially coincide with the reports compiled by emergency response agencies, such as the aforementioned memoir of the firefighters. In that memoir, published by the *Monthly Fire and Civil Defense Magazine*, Andrea Pais wrote:

> What was new this time was all those dead bodies. Seventy days of corpses. What was new this time was sailing on the lake and river, fishing for corpses. They were the dead in piles—the wholesale dead—in the bends of Cadola and Sovérzene. Or they were the single dead—the retail dead—searched for and found along a fifty-kilometer stretch from the disaster site: hidden in the bushes, dug out with bleeding hands from under stones, or soaked, hanging from the beams above.[35]

Death at Vajont made itself known through all those bodies, or perhaps fragments of bodies, that remained imprinted in the memory of survivors and rescuers. But the Vajont dead truly became ghosts, present even when they remained invisible. Alpini officer Italico Cauteruccio describes Longarone as an empty place, "a nothingness extending as far as the eye could see," where there were no dead left to be collected. Yet, those dead were there, down there, everywhere, and the fluctuations of temperature between day and night accelerated their decomposition, releasing, especially at night, "a mephitic stench, somewhere between sweetish and nauseous, [that] fouled the air."[36] And if it wasn't the decomposing corpses, it was the disinfectant scattered everywhere that saturated the area with an unbearable odor. Giuseppe Vazza, like Zanarini, a survivor who traveled through the night of the disaster, remembers that the smell of that night would linger with him for months.[37]

Invisible, vanished dead that would never be found, increasing the controversy over the actual number of victims. For example, Cauteruccio recounts the decision to restore the Cadore road by paving over an area where there were most likely still bodies to be extracted and identified. According to journalist and writer Lucia Vastano, author of a powerful book on the aftermath of Vajont,[38] the bed of the Maè creek would most likely have become a huge mass grave where victims' bodies would be buried under layers of debris. Survivors' accounts would confirm Cauteruccio's testimony by revealing that, in some cases, rescuers had opted to cover the debris rather than dig for the bodies.[39] On the other hand, as one of the survivors interviewed by Vastano recalls, although the fear of an epidemic could justify the urgency of the moment, it certainly could not justify the decision not to resume the excavation and recovery of human remains buried under the rubble. As Lucia Vastano emphasized, he is still waiting to bury at least some part of his grandson, who died and was never found.

The Maè riverbed is then shrouded in another mystery that also refers back to the invisible presence of the Vajont dead. As in a Gothic story, the lunar landscape of the disaster is populated by ghosts and the stories that make them real, even when we do not grasp them. This is exactly how Paolo Rumiz, one of the few writers capable of evoking the ghosts hidden in Italian landscapes, summarizes this history, buried in the bowels of Vajont:

> There is no peace for Longarone. Thirty-nine years after Vajont, a specter still haunts it. That specter is a Dutch, or perhaps German, bus that no one saw that night of October 9, 1963, but which should still be down there in the gravel sarcophagus. A *fata morgana*, many say. An imaginary shadow coming out of the black hole of memory. But in the meantime, digging begins again in the Piave Valley, swept away in four minutes by the wave of the apocalypse.[40]

The story of the ghost bus has been part of Vajont's collective memory from the very beginning. It all starts, once again, with a body—a woman's body, different from the many others that the sea of water, mud, and debris had brought back in the aftermath of the disaster. The tale recounts the discovery of the corpse of a tall, blonde woman with lacquered nails and shaved legs, so strikingly different from the social norms of 1960s Veneto region that it immediately led people to believe she was an outsider.[41] Meanwhile, a bus from a local transportation company was missing from the register. Thus, the story of the ghost bus was born—rumored to be full of foreign tourists, perhaps German or Dutch, swept away by the Vajont wave, possibly returning from Venice. No foreign country ever came forward to claim missing people, and the missing bus—Rumiz reminds us—was found stationary in a parking lot, not loaded with tourists. However, in February 2002, thirty-nine years after the disaster, the ghost bus returned to haunt the Vajont Gorge. Alvise Maso, who was nineteen years old at the time of the disaster, decided to reveal his secret, sharing his version of the events surrounding the ghost bus. A few months after the disaster, in 1964, Alvise was working at a site consolidating the railway bridge over the Maè when his digger struck a large metal object—it was an overturned bus, buried under debris, from which an "unbearable stench" rose. Alvise was instructed to cover up the excavated object and forget the matter, perhaps out of fear that the work might be interrupted.[42]

The fact is that thirty-nine years later, Alvise decided to make peace with his conscience and tell his side of the ghost bus story. After the Vajont Survivors Committee brought the case before the judicial authorities, excavations were carried out, but they yielded no results. The Committee for Survivors still seems convinced of the existence of the ghost bus, despite the failure of the search and especially the absence of any request from a possible foreign country. The truth, perhaps, is that the ghost bus is, more than anything else, a metaphor—a story about the dead who were never found, about the desire to cover the wounds of Vajont too quickly, hiding the debris, both human and nonhuman, as if those lives and that world had never existed. Instead, that story and that world are still there, somewhere. Like a phantom bus, they roam invisibly through the Vajont valleys, trapped in an oblivion that seeks to erase their existence.

2

PROJECTS, PREDICTIONS, OMENS

IMAGINING THE GREAT HYDROELECTRIC MACHINE

> I want to examine the river as a hydraulic machine, as an energy system that, although modified by human actions, retains its natural, spontaneous qualities.[1]

With these words, historian Richard White opens his volume on the Columbia River—a concise yet ambitious book that rethinks the relationship between humans and nature, moving beyond traditional dichotomies such as the artificial versus the natural and the cultural versus the environmental. The metaphor of the river as a great hydraulic machine is equally apt for our history. Claudio Datei, a professor of hydraulic engineering directly involved in the Vajont events, described the Piave Valley in the 1940s and 1950s as "a kind of hydroelectric laboratory"—essentially a vast open-air training ground for a generation of engineers eager to shape the new face of Italy's economic miracle. Like the sirens in *The Odyssey*, the numerous construction sites scattered throughout the valleys attracted these young Italian engineers, driven, as Datei notes, by "that excitement—a sort of drug, perhaps—that came from designing and building big."[2] The siren song, as we know, lures those who heed it into the abyss, leading to inevitable shipwrecks. Since our journey began in the depths, we must now listen to the song of these modern sirens—forces capable of transforming rivers into turbines and valleys into laboratories.

In an Italy in which fossil fuels were scarce, hydropower was indeed a kind of philosopher's stone—wielded not by alchemists but rather by engineers. By the early 1900s, 70 percent of the energy consumed in Italy came from water, making the country the third-largest producer of hydroelectric power worldwide.[3] As historian Giacomo Bonan has pointed out, World War I gave further impetus to the hydroelectrification of Italian waterways.[4] According to James Sievert, in the aftermath of the war, Italy had ninety-one artificial lakes, with an additional forty-one reservoirs still under construction.[5] Technological and scientific progress made it possible to turn rivers into kilowatts, but the hydroelectric option in Italy was certainly not merely about turbines and penstocks. Rather, it was a political project—one might even call it a political ecology project—that established a hierarchical geography of territories, placing some regions in service to the welfare of others, which, by synecdoche, came to be seen as representative of the nation as a whole. The book *La Conquista della Forza* (The Conquest of Power), published in 1905 by Francesco Saverio Nitti,[6] was the manifesto of the hydroelectric shift in Italian politics—an Italian version of the high modernism theorized by American political scientist James Scott that combined two secular religions: faith in technological progress and blind confidence in the central power that imposed it.[7] It was a peculiar form of modernity, then—one that constructed dams like temples and demanded faith rather than critical thought. As is often the case with religions, it exacted its toll in sacrificial victims.

Hydroelectric modernization impacted the entire peninsula, transforming both the Alps and the Apennines. The Piave River basin proved particularly well suited for hydroelectric development. Bonan calculated that between 1925 and 1962, the number of hydroelectric power plants on the Piave River increased from eighteen to fifty-two, with a capacity of approximately 700,000 kWh.[8]

In 1910, paper manufacturer Gustavo Protti sought to harness the power of the Vajont River for his paper mill in Castellavazzo. The initial plan involved building a modest dam just over seven meters high—roughly forty times smaller than the massive structure that still looms over Longarone today. A decade later, in 1926, Carlo Semenza, one of the key figures in this story, arrived at Vajont with initial plans for a comprehensive hydroelectric project for the entire basin.

Semenza, a graduate in hydraulic engineering from the University of Padua, was working for the Società Adriatica di Elettricità, a leading player in Italy's burgeoning hydroelectric industry. At just thirty-three years old, Semenza had a promising career ahead, which would ultimately include prestigious roles such as vice president of the Italian Committee of the International Commission on Large Dams, as well as serving as a United Nations expert on hydroelectric projects in Yugoslavia. His work would extend globally, leaving a legacy of dams built from Japan to Iran and from Brazil to China. Yet, in 1926, Semenza was still a young engineer, navigating the opportunities and challenges of the Piave Valley. Transforming the Piave basin into a vast hydroelectric network was no small feat. Nearly four decades later, in a letter to his mentor, Vincenzo Ferniani, Semenza reflected on the enormity of the challenge: "After so many successful projects jobs and so many constructions, even imposing ones, I am really faced with something that, because of its size, seems to be slipping through my fingers."[9]

By 1961, we were nearing the epilogue, as the anguished tone of this letter suggests. In 1926, however, we were still at the prologue, and the Piave Valley and Vajont Gorge seemed like a vast canvas onto which bold hydraulic projects could be sketched in the name of progress—and, of course, in the interests of investors.

The history of hydroelectric projects in Vajont is long and, at times, complex to follow. Others before me have compiled detailed chronologies that document, one by one, the various proposals through which the Vajont valley was transformed into a hydroelectric engine.[10]

One striking aspect of this "rosary of wet dreams" is the progressive growth of the dam, which, like a plant nourished by the technological imagination of the era, stretches and expands, adding a few meters with each new project. The 1929 plan, designed by Semenza, envisioned a 130-meter-high dam—capable of creating a hydroelectric reservoir holding approximately thirty-four million cubic meters of water.[11] The dam continued to grow throughout the 1930s, with a new design in 1937, and took on the characteristics of the "Great Vajont" by the early 1940s. With the *Grande Vajont*, Semenza aimed to systematize the entire hydroelectric network of the Piave, Maè, Vajont, and Boite rivers, integrating the basins to ensure a consistent and continuous water supply for the Soverzene and Gardona hydroelectric power plants. The goal was to produce approximately eight hundred million kilowatt-hours per year. By this point, the envisioned dam had reached 667 meters above sea level. For those like me, who struggle with mathematics and lack the geometric imagination of engineers, it is worth noting that each small increase in the dam's height corresponded to a significant rise in the reservoir's capacity—from thirty-four million cubic meters in the 1920s to 150 million cubic meters by the 1950s. The Great Vajont project was approved, as Paolini and Vacis recounted,[12] in 1943 in an atmosphere that was surreal, to say the least. In the midst of a rout, with the country divided between Nazi occupation and that of the Allied Powers, the Higher Council of Public Works met in Rome.[13] Although only thirteen out of forty-four members were present, the council approved the project. The fact that it was approved despite the institutional and military crisis of 1943 strongly suggests a close relationship between SADE and government bodies. The issue of illegality due to the lack of a quorum was resolved in 1948, when the new government ratified the decisions made five years earlier. However, the Vajont Dam was not destined to stop at an elevation of 667 meters. In 1957, Semenza, representing SADE, presented the final version of the project. This design proposed a dam 261 meters high, reaching an altitude of 722.5 meters above sea level and with a reservoir capable of holding up to 150 million cubic meters of water.[14] The project had reached its final form: Semenza was now poised to construct the world's tallest arch dam in the Vajont valley, transforming a series of hydroelectric facilities into a sophisticated hydraulic system. After three years of construction, the dam was complete, set amid the rocks of Vajont with its gaze fixed on Longarone and its back turned to Monte Toc. Designed as an engine ready to power progress, it would instead become a cannon, with its aim trained on the valley.

Figure 2.1
View of the reservoir from the Erto and Casso side. Courtesy of the Comitato Sopravvissuti Vajont.

CALCULATED FUTURES

Engineers, at least in the popular imagination, are rarely seen as visionaries—too anchored to their numbers and formulas to let their imaginations take flight. Yet, what else could Carlo Semenza be called if not a visionary? The plans for the Great Vajont—the mathematical calculations, the carefully drawn lines, the formulas for evaluating load capacity and structural integrity—were ultimately an exercise in imagination, envisioning a landscape that heretofore existed only in Semenza's mind and in SADE's blueprints.

The etymology of the word "project" itself evokes this idea of projection into the future. A project, regardless of its nature, is always a vision, sometimes articulated with mathematical equations and plans, as at Vajont, and other times with slogans or images. The Great Vajont, boasting the world's tallest arch dam, a capacity of 150 million cubic meters of water, and millions of lire in capital investment, was far more than just another engineering endeavor.

To bring such a vision to life required not only foresight but also the ability to anticipate and mitigate any unforeseen challenges. Semenza's visionary work in imagining the vast hydraulic system of the Piave, Boite, Maè, and Vajont rivers was extraordinary, but it also required someone capable of interpreting the language of the rocks themselves and predicting their movements. Although, for narrative reasons, I have chosen to present the geological reports on Vajont after illustrating the construction of the dam, this does not mean they were absent throughout the project's development. SADE, the Higher Council of Public Works, the local mayors and their constituents, prosecutors, and lawyers all sought answers about the geological stability of Monte Toc and the possibility of predicting its movements. As Edoardo Semenza, a geologist and the son of Carlo Semenza, explained, prior to Vajont, geological investigations related to dam projects primarily focused on the impoundment zone—the rocky flanks to which the dam was anchored—while largely neglecting the mountain slopes surrounding the reservoir.[15]

Even from these early, highly circumscribed investigations, a defining aspect of this chapter in the Vajont story becomes clear: The mountain spoke in an ambiguous language, offering contradictory messages open to multiple and often discordant interpretations. At the beginning—during the 1920s—there were conflicting opinions about the optimal site for the dam. Professor Hug advocated for the Dogna site, whereas geologist Dal Piaz recommended the Colomber Bridge location.[16]

Dal Piaz prevailed in the debate and, more significantly, became a pivotal figure in interpreting the mountain's language and attempting to predict its movements. In short, Dal Piaz became SADE's geological consultant, and much of the reporting on the state of the Vajont basin's flanks can be attributed to his work.

> The walls of the Vajont Gorge are so steep—often vertical or overhanging—that constructing the road carved into the solid rock of the flanks, connecting the villages of the Piave Valley to those of the Cellina Valley, was an extraordinarily daring and challenging task. One need not elaborate further to grasp that, if there is any location striking in its unique morphological characteristics and seemingly ideal for a dam, it is the Vajont Gorge.[17]

In the tragic case of Vajont, Dal Piaz was notable for his optimism. Even when confronted with less favorable scenarios, the Paduan geologist seemed to maintain an essentially hopeful outlook regarding the stability of Monte Toc and the reservoir's future. Not that Dal Piaz was clueless; his academic biography lists numerous national and international accolades, including his role as director of the Institute of Geology in Padua and his participation in the international commission responsible for drafting the stratigraphic lexicon.[18] However, in the case of Vajont, Dal Piaz is perhaps best remembered for a letter he wrote to his friend Carlo Semenza on February 6, 1957:

> I attempted to draft the statement for the upper Vajont, but I must sincerely confess that it did not turn out well and does not satisfy me. I kindly ask you to send me the text of the version you outlined to me verbally, which I found to be excellent. Additionally, please let me know whether I should include the header of the institution to which it is addressed and whether the date should be current or backdated.[19]

The report that Dal Piaz requested Semenza to draft on his behalf pertained to the 1957 project, which proposed raising the dam to an elevation of more than 722 meters above sea level (261 meters in height). This project had already received approval from the relevant authorities, pending the submission of a detailed geological analysis. Despite efforts to normalize this epistolary exchange, many of us remain puzzled by a geological report completely subservient to the desires and interests of the engineer who was building the dam.[20] This issue would later be highlighted by the Parliamentary Commission of Inquiry on Vajont, which criticized the excessive overlap of roles between geologist Dal Piaz and engineer Carlo Semenza:

> It raises serious reservations and concerns that, on June 11, 1957, Prof. Dal Piaz sent a handwritten geological report to Engineer Semenza, the Vajont planner, just before the Higher Council's vote. In the accompanying note, he stated: "I hope that my writing meets your expectations and that no substantive changes are necessary. Please return the manuscript with your remarks at your convenience, which I will, as always, take into account." This correspondence objectively raises doubts about whether the evaluations and forecasts in the geological reports were guided by objective research criteria or were instead unduly influenced by the "expectations" and preferences of the planner representing the concessionary companies. It is particularly troubling considering that, while Engineer Semenza was a recognized expert in hydroelectric construction, he was not a specialist in geological investigations. This dependency suggests that the reports could have been subject to "substantive modifications" to align with the planner's goals.[21]

Dal Piaz undoubtedly played a central role in the Vajont events, but he was not the only geologist tasked with the challenging job of predicting the behavior of Monte Toc's unstable rock masses. As we will explore below in the section on omens, it was the mountain itself that highlighted its instability, compelling SADE and its planners to anticipate its movements and mitigate potentially catastrophic outcomes.

In 1957, Austrian geologist Professor Leopold Müller entered the scene, invited by engineer Carlo Semenza. Müller identified an unstable mass in the Pian di Toc area and proposed a research plan, which was subsequently entrusted to geologist Edoardo Semenza, the engineer's son. Between 1959 and 1960, Edoardo Semenza conducted geological surveys, initially alone and later with Franco Giudici, culminating in the discovery of the great Vajont paleo-landslide.

The Giudici–Semenza report likely caused considerable concern for Carlo Semenza, who requested that his son submit the findings to Dal Piaz for review and that he temper the report's conclusions. In his recounting of the events, Edoardo denied that his father had asked him to falsify information, asserting that the adjustments were superficial formalities that did not alter the report's substantive findings. According to Edoardo, his father was concerned that the tone of the geological report might alarm someone, especially at SADE.[22] Perhaps others, like me, will wonder whether, in some cases, alarmism is really the worst of evils. The Parliamentary Commission of Inquiry emphasized the exceptional nature of the Giudici–Semenza report: "The significance of the Semenza–Giudici report lies in the fact that, for the first time, and before the landslide of November 4, 1960—the most notable and significant of those that had occurred previously along the Toc walls—a concerning hypothesis was presented regarding the movement of large rock masses in relation to the reservoir's water overflow operations."[23] It was the same commission that recalled how, in his July 1960 report, Dal Piaz had downplayed the risks identified by Semenza and Giudici.[24]

Müller, however, took a diametrically opposed view to Dal Piaz. In his February 1961 report, Müller built on the investigations of Semenza and Giudici to describe the risks looming over the reservoir. For the purposes of my account of the Vajont disaster—and especially considering my abysmal ignorance in geology—I will not delve into the technical details of the different reports. Instead, I would say that, in contrast to the optimism of Dal Piaz and other experts surveyed by SADE, Edoardo Semenza, Giudici, and Müller identified the huge risk of the Toc landslide, highlighting the dangers it posed to the hydroelectric basin. In his February 1961 report, Müller spoke of a landslide involving about two hundred million cubic meters of rock—a nearly perfect prediction of the actual size of the landslide—which, in his view, could not be artificially halted, implicitly suggesting, according to the Parliamentary Commission of Inquiry, the complete abandonment of the project.[25] On the other hand, Müller himself linked the triggering of the landslide to the operations of the reservoir: "Before the existence of the reservoir, these movements would have been only a few centimeters per year, whereas with the dam, they increased to several decimeters. Consequently, an accelerating and amplifying effect can be attributed to the reservoir."[26]

Yet, Vajont seems like the perfect case to understand what post-normal science, as theorized by Silvio Funtowicz and Jerome Ravetz more than twenty-five years ago, is all about. For Funtowicz and Ravetz, post-normal science comes into play when uncertainty is high, opinions are discordant, and affected communities demand involvement in decision-making. This almost mirrors the Vajont disaster, with its conflicting geological reports and citizens concerned by ominous signs. In such cases, Funtowicz and Ravetz advocate for an extended peer review involving the community in the construction and legitimization of knowledge.[27]

What might have happened to Vajont if a post-normal approach had been adopted, where the concerns of local communities were listened to and their empirical knowledge considered—or, at the very least, if a strong precautionary principle had been applied in the face of divergent expert opinions? Unfortunately, none of this transpired at Vajont. Instead, expert disagreements consistently leaned toward the more optimistic assessments, while the concerns of the local population were systematically ignored, if not ridiculed. The person who most embodied this ongoing attempt to mediate between these diametrically opposed viewpoints was Francesco Penta, a professor of applied geology at Sapienza University and a member of the testing commission. In a 1960 report, Penta entertained both possibilities: the Vajont landslide might occur as either a slow and superficial landslide or a sudden and massive geological event. He didn't believe there was sufficient evidence to firmly support either hypothesis, instead recommending further investigation. This interpretation of the situation and his recommendation for continued monitoring were endorsed by the testing commission, which, in December 1960, concluded that there was no immediate risk to the local population.[28] A year later, in October 1961, the monitoring plan does not appear to have yielded any certainty. Penta continued to contemplate the possibility of a sudden detachment of a large rock mass triggered by the filling of the reservoir, suggesting that "means should be sought to avoid, if possible, major landslides and to limit their effects on the lake and ancillary works."[29] How the filling of the reservoir could have been authorized on such fragile grounds remains a mystery. One can imagine that, in Penta's hypothesis—shared by Dal Piaz—the potential risks to local people were not even considered. However, the precautionary principle might have suggested a different course of action.

What is certain is that, despite moving forward with the reservoir project, predictions about the geological future of the basin must have raised serious concerns, requiring significant interventions. Among these was the construction of a bypass tunnel, designed in case the basin was split in two by a landslide.

Although armed with different knowledge, interests, and sources of power, everyone involved found themselves scrutinizing the slopes of Monte Toc, almost trying to predict what might happen. Nature, as we know, however, is not easily interpreted; it often appears confusing, uncertain, and unpredictable.

Yet, one could argue that it might not be entirely correct to speak of nature at Vajont, where the hydroelectric company had transformed an entire basin into a hydraulic machine. Indeed, SADE entrusted the "Ettore Scimemi" Hydraulic Modeling Center in Nove with the task of bringing the valley into the laboratory and testing the effects of a potential landslide on the reservoir. The entire Vajont basin was reproduced at a scale of 1:200, and under the direction of Professor Augusto Ghetti from the University of Padua, twenty-two experiments were conducted to better understand the dynamics triggered by a landslide within the reservoir.[30]

The scientific debate surrounding the Ghetti model, like everything related to Vajont, is highly intricate and perhaps difficult for nonexperts to follow. Some, including Marco Paolini, have argued that by using gravel to simulate the landslide, the model downplayed the severity of its effects on the basin. Indeed, years after the disaster, in 1967, a French hydraulics laboratory repeated the experiments using compacted material instead of gravel and obtained different results.[31]

The time of the landslide's fall used in the Ghetti model would become another point of contention. In reality, the colossal rock mass plunged into the lake in approximately twenty-nine seconds—"that is, not 8 seconds, but more than twenty times the minimum time indicated by SADE when commissioning the study."[32] At the trial, Ghetti would repeatedly emphasize that it was not his role to question the basic data—"that is, volume, configuration of the landslide, and time of the landslide's fall"—as provided by the client.[33] One passage in the final report on the Novi experiments is particularly controversial: "By decreasing the elevation of the initial reservoir, the effects of overflow and spillage are rapidly reduced, and already, at an elevation of 700 m a.s.l. [above sea level], the reservoir can be considered absolutely safe even in the event of the most catastrophic foreseeable landslide."[34]

It was based on this categorical statement of absolute certainty that Professor Augusto Ghetti was brought to trial. As reconstructed by his student and defense consultant, Professor Datei, the prosecution's experts argued that the hydraulic model was flawed and inadequate for providing reliable predictions. Therefore, they held Ghetti responsible for the statement regarding the seven-hundred-meter threshold.[35] The defense was able to demonstrate that the model was correct, whereas it was the data provided by SADE that had failed. Despite the seven-hundred-meter limit, SADE had still requested and obtained permission to raise the reservoir level to 715 meters. At the same time, however, it had rejected

Professor Ghetti's recommendation to continue and expand research across the entire Vajont.[36] Rightly, Datei wonders whether the authorities would have granted this authorization if they had been informed of the recommendations of the hydraulic models center.[37] In the end, Ghetti was acquitted by a full verdict, but I am not sure that this acquittal resolves all the questions left open in the history of the Vajont hydraulic model. Beyond any consideration of criminally punishable offenses, the criticism raised by the investigating judge regarding Professor Ghetti's lack of prudence seems reasonable to me. Referring to the statement about the absolute safety of the seven-hundred-meter elevation, the investigating judge wrote: "Faced with the unknowns that had made experimental research difficult, the manner in which the tests were conducted, and the lack of guarantees that a model never before used could provide, Ghetti should have exercised more caution in his language."[38]

MESSAGES FROM THE UNDERGROUND

It wasn't just technical reports and piezometric wells that foretold the mountain's future. Many people were scrutinizing the rocks and trees of the Toc, listening to every faint whisper and recording its movements. As in any classical tragedy worth its name, at Vajont, the gods—or perhaps nature—did not fail to send their omens. The history of the dam is filled with signs, messages invisible to the powerful but unmistakable to the valley's inhabitants. Many remember the rumbles and tremors, like small earthquakes, that accompanied the filling and draining phases of the reservoir. When journalist Lucia Vastano asked Irene De Lorenzi about the night of the disaster, De Lorenzi promptly replied: "It is not possible to start from the end. My fear is much older. It all started with the construction of the dam because that's when the earthquakes and landslides began."[39]

Irene recounted the 1957 earthquake, describing stones "splashing out of the walls" and then an incident that occurred on the Toc slopes in 1960. While carrying a basket, a woman was swallowed by a sudden chasm that opened in the pasture—a kind of revelation that, as Irene recalled, made it clear to those present that "the mountain was sliding down." The almost pastoral setting of this scene should not mislead, however. Irene's family exemplified the strange hybrid of modernity and rural life that characterized Vajont. Irene's husband was a worker employed at the dam. He had repeatedly tried to warn management about the ominous signs coming from the mountain, but, as Irene lamented, his efforts were in vain. No one wanted to hear his accounts of roars, cracked walls, or leaning trees along the reservoir's slopes. Instead of heeding his warnings, the dam's leaders mocked him, repeating a phrase that encapsulates the entire Vajont tragedy: "You are mountain people. What could you possibly understand?" Fortunately, Irene's husband chose to listen to the Toc rather than the site bosses. He quit his job, saving himself from the certain death he would have faced had he stayed at the dam.

Irene's husband was not alone in recognizing the mountain's warnings. Giuseppe Vazza, one of the survivors introduced in the first chapter, recalled the alarming events around the dam in the weeks leading up to the disaster. In September, during a delivery to the site's canteen, a massive landslide plunged into the lake, raising a wave as tall as a bell tower. The wave struck the reservoir's banks and violently shook the canteen's kitchen shack, although it caused no direct harm.[40] Like Irene's husband, the cook in this story reportedly decided to quit her job after this incident; the mountain clearly instilled fear in those willing to listen.[41] Not content with this close encounter with the power of the landslide, Vazza recounted his own exploration of the basin's slopes:

> Driven by curiosity, while my mother and wife were picking mushrooms, I climbed up the steep slope of Mount Toc. After about twenty minutes of climbing, I saw it—a terrifying scene! The mountain beneath my feet was separated by almost two meters from the rest of the slope. In front of me was a chasm, not very deep but stretching as far as the eye could see down the mountain. Inside that crevasse, there was a slow overturning of earth. Stones were submerged in a kind of slime that appeared liquid. I noticed plants tilting and falling. I was deeply shaken and came down almost running to where my mother and wife were standing.[42]

These were certainly individual experiences, but they were always amplified by the dense network of community relationships typical of that social and historical context. Vazza, the village butcher, did not hesitate to share the discoveries he had made on the Toc with his customers—not that the valley residents needed their butcher's tales to understand that the situation was becoming extremely dangerous.[43] According to Vazza, in Erto, "not a day went by without landslides, small earthquake tremors, or cracks and plaster detaching from the walls of older houses."[44] Earthquake rumblings and tremors were the order of the day according to the testimony of Longarone parish priest Pietro Betz.[45] During a television program dedicated to Vajont, a witness recounted how the movement of the landslide was visible to the naked eye through the tilting of the rows of fir and larch trees. Additionally, the landslides were clearly perceptible—especially to someone like him, who rode a Lambretta—due to the constant potholes in the road, which

SADE continually covered.[46] The Ministry of Public Works report includes the testimony of a truck driver, Antonio Savi, who stated that the road between Erto and Casso and the Toc was becoming increasingly unsafe.[47] According to municipal guard, Matteo Corona, SADE was paying workers to "hide" the potholes that continually appeared on the road—an operation that, he said, was complicating efforts to monitor the risk.[48] In a letter sent to all relevant authorities and ENEL-SADE, the mayor of Erto and Casso succinctly captured the state of apprehension gripping the population due to these omens: "The people of Erto and Casso are living in a state of constant apprehension and alarm. While some may downplay these issues, for the people of Erto, they directly affect the security of their lives and property. This administration once again expresses its concerns for the safety of the people and the village."[49]

As often happens in situations like this, it was not only humans but also—indeed, especially—nonhumans that warned of the impending danger. One SADE worker, Renato Trevisol, recounted how processions of shepherds came to the dam site to inform the company that their animals were too frightened to eat the pasture grass.[50] Postal employee Maria Pia Bassetto[51] and Giuseppe Vazza again told of nervous animals fleeing the woods.[52] Gervasia Mazzucco spoke of an atmosphere of fear that enveloped both humans and animals—a fear so deep that it even penetrated people's dreams. Just a week before October 9, Gervasia's mother had dreamed of drowning.[53]

Cracks in the houses, potholes in the streets, roars and tremors, tilted trees, and nervous animals—this was everyday life at Vajont, where an entire mountain was set in motion on the banks of an immense artificial lake. At the trial, one of the plaintiffs' lawyers, Giorgio Tosi, ironically described the clumsy attempt to downplay and normalize the omens coming from the Toc. Speaking about the report of the Parliamentary Commission of Inquiry, Tosi focused on the use of the adverb "imperceptibly":

> At one point, we read that the Toc landslide of 250 million cubic meters had moved, from 1960 to 1963, "imperceptibly." Now, we all know from the court documents that, in three years, this huge mountain had shifted, had moved (horizontally) toward the lake by 4 meters. "Imperceptibly" is a very pretty, I would say graceful, precious adverb; and like all precious things, it must have cost a lot. I would not dare to call ignorant those who humorously used the adverb "imperceptibly" for the 4-meter displacement of a huge mountain, as if the Toc, a modern Narcissus, had wanted to mirror himself in the lake and demurely, playing hide-and-seek with the clouds and taking advantage of the night's darkness, had tried to hide this "special friendship" of his with an imperceptible 4-meter displacement. . . . All of this was macroscopically perceptible: the great fissure, the displacement of the strongholds, the tilting of the trees, the partial landslides, the creaks, the roars, the decay of the elastic modulus, the alarm of the population, and the very concerns.[54]

Some portents conflict even more dramatically with the rhetoric of imperceptibility. On March 22, 1959, a massive landslide of three million cubic meters of rock plunged into the Pontesei reservoir, also part of the Great Vajont hydroelectric system, killing one of the workers, Arcangelo Tiziani. His body, swept away by the landslide and drawn into the reservoir, was never recovered—an invisible death, just as, in fact, the Pontesei landslide itself was almost invisible, ignored both by SADE technicians and by the controlling authorities.[55] The following year, on November 4, 1960, it was the turn of the

Toc, with a landslide of seven hundred thousand cubic meters of rock that detached from the left bank and crashed into the basin, raising a wave ten meters high. More than an omen, the November 1960 landslide seemed like a warning. Even SADE's engineers were deeply shaken, as evidenced by a letter from Carlo Semenza to his teacher and friend, engineer Ferniani, dated April 1961: "I won't hide from you that the problem of these landslides has been bothering me for months; things are probably bigger than us, and there are no adequate practical measures."[56]

As I have already reported, SADE decided to proceed with the construction of a bypass tunnel that would allow the basin to be used and prevent it from overflowing in the event that a large amount of rock occupied part of the lake. More generally, following the recommendations of geologist Penta, it was decided to deepen geological research on the landslide and constantly monitor the mountain. In fact, as also pointed out by the Parliamentary Commission of Inquiry, the geological research on the Toc was already quite thorough and would have required drastic risk prevention measures that were not implemented—not even a warning system for the communities in the area was set up. A network of powerful beacons illuminated the slopes of the Toc, while piezometric wells monitored the movement of the rock masses. The landscape of Vajont was certainly far from reassuring. This is why, in contrast to SADE's sophisticated equipment, the people of the valley relied on their own DIY monitoring systems—scrutinizing tree lines, cracks in walls, and the anxious behavior of animals, and even employing the most ingenious methods in the final days, such as the one recounted by municipal guard Matteo Corona: "I was instructed to go to Mt. Toc to inspect the cracks that had already formed in the shape of an 'M' up there. I made spies with a billhook that I had in my backpack; I made sticks upstream and downstream and marked numbers on them [unintelligible]. According to the movement of the material, I would report back to my mayor."[57]

Scientists versus mountain people, wooden sticks versus piezometric wells, rumblings from the bowels of the Toc versus official reports. In the conflicts of knowledge at Vajont, there was no match. When a reporter from the public television asked why they hadn't all run away earlier if they knew of the danger, one survivor replied, "With the hope that they knew more than we did."[58]

Those mountain people, in short, knew—if only someone had been willing to listen. Just as Tina Merlin did.

3

THE VAJONT CASES

CASES WON, CASES LOST

On November 30, 1960, Tina Merlin, journalist, and Orazio Pizzigoni, editor of the newspaper *L'Unità*, were acquitted by the Milan court of the charge of spreading "false and tendentious news aimed at disturbing public order." The trial against Tina Merlin was the first of the Vajont trials and, paradoxically, ended by siding with the great accuser of SADE—the Cassandra of Vajont as she was often called.[1] Unfortunately, some cases are lost even when they are won in court.

The subject of the court dispute was an article by Merlin, published in *L'Unità* on May 5, 1959, with the unequivocal headline "SADE lords it over, but mountain people defend themselves." The article reported on the formation of a consortium among the inhabitants of the Erto valley to protect themselves from the abuses of the large hydroelectric corporation. Although much of the article focused on the expropriations and restrictions on movement imposed by SADE, the lawsuit mainly targeted the concerns and risks posed by the dam. The subpoena, in fact, held Tina Merlin and Orazio Pizzigoni accountable for spreading news about the "existence of a serious danger" to the communities bordering the reservoir.[2]

Recounting the founding assembly of the consortium, Tina Merlin wrote:

> Families directly affected by the threat of expropriation or the encroachment on their property by SADE, along with many other mountain residents, spoke out. They saw in the company's selfishness and the government's inertia a grave threat to the very existence of the village, with a reservoir of 150 million cubic meters of water being built behind it. This reservoir could eventually erode the landslide-prone soil, potentially causing houses to sink into the lake.[3]

A few experts pointed out the errors made by Merlin and the villagers in their predictions regarding the risks posed by the dam. For my part, I believe it is irrelevant to highlight the lack of expertise in geology and engineering of the journalist and the local inhabitants, as this is a truism; in fact, it would be strange if this were not the case. On the contrary, it seems to me that both the citizens gathered in the assembly and the journalist from *L'Unità* were carrying out their duties very well, raising their concerns and denouncing the authorities' lack of attention to their demands. It was not their job to solve the complicated geological and technological puzzle of Vajont. That they understood the geological dynamics of Monte Toc was by no means enough, since the massacre was not prevented, and no one forced SADE to abandon the plant—not even those with the academic qualifications and expertise appropriate to the case.

The issue of the risks triggered by the dam was undoubtedly the central concern around which the consortium for the defense of the Erto valley was formed. The journalist Renato Zanolli connected the Pontesi landslide of March 1959 with the consortium's founding assembly, convened a few months later.[4]

In the aftermath of the disaster, Merlin herself recalled the 1959 assembly as a revelatory moment that unequivocally demonstrated the local communities' awareness of the risk:

> I had attended an assembly of farmers from Erto and Casso. All the heads of families had formed a consortium to defend the Erto valley because they were convinced that if the power company built the reservoir, it would pose a serious danger to them. This was because the village of Erto was built on the site of an old landslide that had fallen from a nearby mountain hundreds of years ago.[5]

The trial against Tina Merlin and *L'Unità* became an opportunity to bring the concerns of valley residents into an institutional forum. The defense called three citizens of Erto—Celeste Martinelli, Pietro Della Putta, and Osvaldo Carrara—as witnesses to demonstrate how widespread the apprehensions about the Vajont Dam were, independent of the incriminating article.[6] In their testimony, Martinelli and Della Putta explicitly referred to the Pontesei landslide as a pivotal event that heightened awareness of the risks associated with the Vajont basin. Della Putta added details about the anxious state of the people of Erto, fueled by cracks in the walls of their houses, continuous seismic tremors, and visible fissures on the slopes of the Toc—those ominous signs we encountered in the previous chapter.

In the end, as mentioned earlier, Merlin and Pizzigoni were fully acquitted. For the judges in Milan, Tina Merlin had simply exercised her "right to report news, informing the public of entirely legitimate facts and criticisms."[7] The ruling not only rejected the claim that the news published by *L'Unità* was false or tendentious but also dismantled the accusation that it aimed to disturb public order. As stated in the ruling:

> The news reported in Merlin's article is devoid of any causal effect in relation to that event, given that the environment and community to which it refers—and in which it was disseminated—was already deeply disturbed and concerned about it due to circumstances and fears that were entirely unrelated, independent, and prior to the defendants' activity. In this respect, therefore, the conduct of Merlin and Pizzigoni introduced nothing new and had no causal effect or capacity to disturb public order, which was already present.[8]

Later, when the disaster had already unfolded, Tina Merlin bitterly remarked, "Had I really managed to upset the SADE order, we would not be here today mourning our dead and cursing those responsible!"[9]

Tina Merlin was, indeed, a passionate advocate for the lost cause of Vajont, extending her efforts far beyond the Milan trial. Historian Adriana Lotto has written an insightful biography of Tina Merlin, situating her fight at Vajont within the broader context of her life as a communist journalist and activist.[10] It is difficult to fully grasp Tina Merlin's relentless denunciation of the events at Vajont without knowing at least a little about her life story. She was a poor mountain girl who left elementary school to work as a domestic servant for wealthy families in Milan. Having experienced marginalization firsthand, she became a partisan and a communist, driven by a profound thirst for justice and a determination to challenge inequality and oppression. For Tina, Vajont was fundamentally this: the story of marginalized mountain communities oppressed by a powerful hydroelectric corporation and a complicit, subservient political system. Like me, Tina did not view Vajont as an aberration of the system, a mistake, or even an unfortunate accident. This is not to suggest that anyone deliberately orchestrated a massacre by building a massive hydroelectric reservoir on the edge of a landslide. The issue lies elsewhere: A socioecological system entirely subordinated to profit inevitably generates risks for the powerless—risks that it systematically underestimates because its functioning prioritizes the production of value over the very reproduction of life itself. I have spoken elsewhere about the "wasting relationships" that create socioecological dumping grounds—where the havens of the privileged always depend on creating hells for the many, who are excluded.[11]

Vajont, like all major projects of the hypermodernist era, is a product of the simplification of socioecological relations—a phenomenon described by American scholar James Scott in his book *Seeing Like a State*. To achieve this transformation, one must view reality as the state sees it—and, I would add, as capital sees it—reducing the complexity of socioecological relations into something readable and classifiable. However, the story of Vajont vividly demonstrates that imposing the perspective of the state and capital on such relations only subjects them to that vision without necessarily ensuring a genuine understanding of how they function.

Tina Merlin, on the other hand, did not view Vajont through the lens of the state or corporations. As a communist, she approached it with a nuanced awareness of the complexity of a landscape where power, knowledge, technology, collective rights, and private profits were deeply intertwined. She expressed this perspective with a clarity more effective than my own: When French journalists asked how she knew about the risks of the dam, she simply said it was enough to listen to the mountain people and trust their concerns.

Tina Merlin's Vajont exemplifies what it means to examine socioecological relations without ignoring or naturalizing power dynamics, instead choosing a perspective rooted in the experiences of subalterns—those who are dismissed because their knowledge lacks institutional legitimacy. Reflecting on Merlin, geologist Edoardo Semenza, the discoverer of the Vajont paleo-landslide, remarked that "the author's political vision led her from the beginning to view the whole affair with the prejudice characteristic of her side."[12] I suppose the corollary of this statement is the presumed existence of an objective approach—one free from bias, immune to the influence of power and economic interests. In other words, it evokes the so-called god trick, which philosopher Donna Haraway unmasked years ago for what it truly is: a rhetorical device that conceals the positionality of the observer.[13] On the contrary, the case of Vajont demonstrates how politicizing the problem could have helped address it effectively. Taking the side of the subalterns, recognizing the asymmetries of power and resources at play, listening to their concerns, and exercising a systematic distrust of those who consistently defend the interests of the powerful are all strategies that would have aligned with the precautionary principle, perhaps leading to further research and stricter controls.

As I have argued previously, it is not that Tina and the communists "politicized" Vajont; that landscape, shaped by turbines and kilowatt-hours, was already a political landscape, shaped by layers of power and power relations that determined its fate. However, in mainstream narratives, "politicization" is often invoked only when subalterns attempt to resist or even speak out. In this framing, political agendas—always those of marginalized people—are seen as preventing an objective view of the facts, which is, of course, the view of those in power. Reaffirming the political nature of socioecological issues is, after all, the core mission of political ecology as a field of research and of environmental justice movements as a field of action.

Unsurprisingly, Tina began her 1983 book on Vajont with a forceful denunciation of the complicity between science and power in producing the disaster:

> It will remain a testament to the everlasting shame of science and politics—a union between science and politics that, twenty years ago, closely tied nearly all distinguished academics to economic power, in this case, the SADE electric monopoly. This monopoly, in turn, used political power—embodied by the Christian Democrats—to carry out large projects for so-called public utility, from which it made, or expected to make, enormous profits.[14]

"Power" is the key word in Tina Merlin's reading of the case of Vajont: the power of SADE, which was able to impose its decisions and neutralize any safety measures; the power of the state, which supported the large hydroelectric enterprise while ignoring calls for precaution; and the power of official science, which, at best, was incapable of prioritizing the safety of citizens over other considerations—and this, in fact, seems to me the most benevolent possible interpretation.

After her 1959 acquittal, Merlin remained concerned about Vajont. On October 8, 1960, in the aftermath of the landslide that fell into the basin, she wrote in *L'Unità*:

> The Erto Reservoir . . . is already beginning to cause disasters. A huge landslide rushed into the lake, breaking off from the land on the left bank in the locality of Toc, just above the great Vajont dam. . . . One was, therefore, right in gathering the concerns of the population and recalling previous experiences in Vallesella and Forno di Zoldo, to alert about the existence of a clear danger posed by the formation of the lake. And the danger has become more and more imminent. At the site of the landslide, the ground continues to give way; there is an impressive noise of earth and stones continually falling. The wide cracks in the ground, encompassing an area of several kilometers, certainly cannot be reassuring.[15]

The contrast between SADE's Vajont and that of the locals becomes a vivid image in Merlin's article when the journalist describes the arrival of the company's "bigwigs" in their "very luxurious cars." All of them inspected the dam and the mountain, but they were ready to flee if any villager approached asking for clarification. Quoting the citizens of Erto, Merlin concluded, "They only care about their lake; they really don't care about us." The following year, on February 21, *L'Unità* published an article by Merlin that would become the grim prophecy—albeit wrong in its details—of the 1963 disaster. The journalist warned of the risk that a massive landslide, involving fifty million cubic meters of rock, could plunge into the reservoir, causing catastrophic damage to the nearby municipalities. As we now know, the landslide was five times larger than the one feared by the journalist, who nevertheless pinpointed a key issue in the dynamics of the disaster: the uncertainty surrounding the timing of the landslide's entry into the reservoir.[16] Tina seemed to underestimate the risks to Longarone—which she feared might only be overwhelmed by the dam's collapse—and instead focused on the dangers to Erto and Casso. But hers were certainly not academic articles destined for a geology journal; others, with the necessary resources, authority, and knowledge, ought to have taken up that cry of denunciation and done what was necessary. Tina Merlin's denunciation, as well as the concerns of the mountain people she had chosen to listen to, fell on deaf ears. On October 10, 1963, from Ponte nelle Alpi, while trying to force a roadblock and reach Longarone, Tina attempted to inform the readers of *L'Unità* about the disaster:

> I am in Ponte nelle Alpi: the road is blocked by police officers and soldiers. It is not possible to pass. Only ambulances, police,

> and army vehicles can pass the roadblock and advance toward Longarone, the town of two thousand inhabitants submerged in the night by the avalanche of water that swept over it after the dam on the Vajont gave way. . . . An eyewitness brought the following dramatic news from Longarone: three-quarters of the village was swept away. The appearance is chilling; not a groan can be heard, it looks like an immense cemetery. Many dozens of people, whole families, are missing. The hamlets around Longarone have also been hit by the huge mass of water: Pirago has been completely destroyed, Villanova and Faè half-destroyed, Codissago much damaged.[17]

According to Toni Sirena, Tina's son and a journalist himself, his mother never fully recovered from the bitterness of failing to prevent the Vajont massacre.[18] As Tina herself wrote a few days after the disaster, "I am writing these lines with a heart gripped by remorse for not having done more to induce the people of these lands to rebel against the deadly threat that has now become a tragic reality."[19] Her commitment to Vajont was not that of a passing journalist, ready to chase the news of the moment; her personal and political biography—which, for those like Tina and me, are the same thing, of course—tied her to that event. She remained connected to Vajont her entire life, following it as a journalist and writer well beyond 1963. Toni Sirena recalls his mother's annoyance at the ease with which even her political side distanced itself from Vajont. Sirena mentions his mother's refusal to accept a plaque offered by the municipality of Longarone to all journalists who had covered the disaster—without distinguishing between those who had pointed out the responsible parties and those who had criticized the communists for doing so. Again, Sirena recounts his mother's annoyance when the National Association of Italian Partisans of Belluno invited

Benigno Zaccagnini to an event, forgetting his role as minister of public works in the Vajont events.[20]

Testifying to this unwavering loyalty to the (both won and lost) Vajont cause is the book published twenty years after the massacre, in 1983, in which, on the final pages, Tina asks herself: "Young people, what do they know about SADE, the struggles of their parents, the abuse of power by the powerful against a community that only asked to live and work in peace?"[21]

I will explore the memory of Vajont more extensively in the final chapter of this book. For now, it is enough to reiterate, along with Tina Merlin, that although remembering is essential, it is equally important to recognize that memory can take many forms. For Tina, remembering went hand in hand with justice; it was not enough to mourn the victims without holding those responsible for their deaths accountable.

IN AGRAMANTE'S CAMP

I do not know if Tina and the Vajont survivors truly believed they would find justice in courtrooms. Perhaps the plaintiffs' lawyer, Sandro Canestrini, spoke for all of them when he declared forcefully in his closing argument on September 23, 1969:

> To the defense attorneys and their protégés, we express our discomfort at being forced to fight in Agramante's camp (and not just in a legal one) within the system. The indictment—guilt with foresight, a concept so strange and with such blurred boundaries—is the one you prefer, instead of an indictment for voluntary manslaughter with malice aforethought. The penal code and jurisprudence are what you favor: a code ruthless about petty theft yet indulgent with manslaughter, a code (and remnant

of fascism) that must regulate this trial, which is itself a cry of revolt.

A venue that suits you, secured through your countless efforts to invoke and obtain legitimate suspicions; proceedings timed to your liking, skirting the edge of the statute of limitations; judges, whose personal expertise and impartiality we won't even question, because they are so obviously products of your world and your society; and all of this, spiced up with a formidable team of defense attorneys cherry-picked from the elite, operating in a political climate that works entirely in your favor.[22]

Nonetheless, it was in that Agramante's camp[23] of courtrooms—on foreign soil, so to speak, and not only because the trial was transferred to L'Aquila due to legitimate suspicion[24]—that the battle for justice in the wake of the massacre was fought. The criminal outcome—leaving aside the civil proceedings—of this Vajont judicial case is well known: After three levels of trial, the Supreme Court found Biadene, an engineer at SADE, and Sensidoni, a member of the dam inspection commission and inspector general of the civil engineers, guilty. The former was sentenced to five years and the latter to three years and eight months, of which three years were pardoned. All other defendants were acquitted, either for not having committed the act (Tonini, Marin, and Ghetti) or for lack of sufficient evidence (Frosini and Violin).[25]

Although the sentences were decidedly mild—considering that, in the first instance, the prosecution had requested twenty-one years for all the defendants except Violin, for whom it had requested nine years—the Supreme Court's ruling was considered a success by many survivors. This was because it reestablished the principle of the foreseeability of the catastrophe, which had been denied at previous levels of justice.

A separate issue would be to examine the prison conditions and privileges enjoyed by Biadene and Sensidoni during their incarceration. Building upon journalistic sources from the time, the Wikipedia biographical entry on Biadene describes his prison regime as including meals paid for by ENEL,[26] a single cell, an inmate acting as his attendant, and a schedule divided between the library and consulting work on prison modernization.

Although this account may leave readers like me dumbfounded—or perhaps even more convinced that justice does not coincide with law—the overall tone of this brief biographical entry is far from guilt-ridden. One reads on Wikipedia:

> With his simplicity, and above all with his seriousness, he had managed to make himself loved by everyone: by the inmates, to whom he gave lessons in mathematics, Italian and Latin, and by the prison guards, who helped him organize the library, which was significantly enriched at the time with volumes from Biadene's personal collection.[27]

It was no coincidence that the radical left-wing daily *Lotta Continua*, reporting on Biadene's prison regime, included news of a young Sicilian man who had been sentenced to two years in prison for stealing fifty-five lire.[28] Sensidoni never set foot in prison, spending his imprisonment in a Roman clinic. The judicial proceedings regarding the criminal trial for the Vajont massacre ends here, with a rather modest sentence handed down by the Supreme Court, a punishment that seems to betray its class character and its limitations. As Canestrini used to say, the system is inflexible with those who steal an apple but rather sympathetic to the powerful.[29]

Journalist Fiorello Zangrando, who had covered the Vajont trial for *Il Gazzettino di Belluno*, recalls that at the reading of the sentence, someone in the courtroom shouted, "The law applies equally to all, but not the rich."[30] However, the denouement of the case by no means exhausts the interest in the long judicial history of Vajont; in fact, it is more accurate to say that it is the trial itself, rather than its final result, that holds the greatest significance.

The Vajont judicial proceedings began just hours after the catastrophe, when public prosecutor Arcangelo Mandarino initiated an investigation against unknown persons, ordering the seizure of all documents related to the dam and gathering the first testimonies of survivors. This marked the beginning of what would become the vast Vajont archives: a collection spanning fifty linear meters and containing 256 folders. These archives were discovered in the basement of the L'Aquila state archives after the 2009 earthquake and are now housed at the State Archives in Belluno.[31]

The trial was transferred to L'Aquila at the request of the defendants, who had asked for and obtained a transfer from the "natural" venue in Belluno on the grounds of legitimate suspicion. They argued that the general climate resulting from the disaster would prevent the trial from being conducted smoothly and impartially. In his argument, lawyer Canestrini described the application of legitimate suspicion as yet another injustice perpetrated against the victims of Vajont, who were once again "pointed to as a reason for disorder to be repressed and punished."[32]

The criminal trial was not the only space to seek the truth about the disaster: There was also a Parliamentary Commission of Inquiry, as well as investigations by ENEL and the Ministry of Public Works. Much of the documentation produced in these forums later found its way into the trial records, which form the extensive Vajont archives.

The Vajont trial, and more generally all investigations into the disaster, centered on the question of the predictability of the landslide and its consequences. After all, public debate also focused on this issue, with journalists of the caliber of Dino Buzzati and Giorgio Bocca[33] firmly siding with those who explained the disaster as an unpredictable natural event.[34] Clearly, if that was indeed the case, no one could be blamed; all that remained was to curse the mountain and mourn the victims. However, this explanation conflicted with Tina Merlin's articles on the risks dating back to 1959, with the many warnings described in the previous chapter and even with the technical reports from some of the experts who had warned about the geological fragility of the Toc. It was for this reason that among the first moves of the Belluno prosecutor's office was the appointment of a commission of experts[35] with the task of answering eleven questions that were perhaps, at bottom, only one: Was the Monte Toc landslide foreseeable in the form in which it occurred on the evening of October 9, 1963?

After eighteen months of work—despite the prosecutor's office demanding that the report be completed within three months—the commission of experts adopted the thesis that the disaster was unpredictable, effectively absolving all those responsible. As Eleonora Pierobon wrote in her comprehensive master's thesis on the disaster, the prosecutor's experts argued that "the factors that determined the catastrophe had to be sought outside the conduct of the men who had been in charge of Vajont."[36] This version, however, did not convince the investigating judge, Mario Fabbri, who appointed another panel of experts, largely composed of foreign specialists, with the exception of geologist Floriano Calvino.[37] If the new panel of experts had no doubts about the predictability of the landslide, it was up to the investigators to prove that

SADE was aware of the risks and that the authorities, too busy pleasing the hydroelectric company, failed to carry out their regulatory duties.

Certainly, the Vajont trial is a peculiar case; reading the thousands of documents—or even just a fraction of them, as I did—it feels as though it moves between a geology or engineering laboratory and a popular assembly. Technical and scientific knowledge obviously played a key role, but broader political considerations also often emerged in a trial that, despite the investigators' intentions, seemed to indict an entire system of production and its dangerous relationships with power and science.

Giorgio Tosi, one of the plaintiffs' lawyer alongside the aforementioned Canestrini and, like him, a former partisan and militant communist,[38] emphasized the contradiction between a trial that was technically for manslaughter—in his words, similar to what might happen in a car accident—and the systemic significance of what happened at Vajont and its translation into the courtroom. Speaking to the judges in the L'Aquila court, Tosi stated: "Some have argued that the Vajont disaster is a consequential, linear, and logical outcome of the system; it is proof that the system is flawed. Therefore, it is the system that must be judged and condemned. Others, however, argue that the Vajont disaster is a degeneration of the system."[39]

In short, it was about identifying personal responsibility and punishing accordingly, without losing sight of the power structures that had allowed and supported those criminal choices. Similarly, the trial clearly investigated the construction and legitimization of scientific and technical knowledge, as well as the difficulties involved in predicting the interactions between large hydroelectric infrastructures and geological structures. However, the trial was not, and could not be, turned into an academic dispute over these issues. It was Tosi again who made this point very clear:

> We must not allow this temptation of meta-legal speculations to take the reins, but neither should we repress or stifle it. . . . If I were to turn my statement into a rallying cry, I would certainly be a bad lawyer and also a bad politician. However, if this is true for politics and sociological inquiry, it is equally true for technique and science. . . . So much so that, if I were to give a dissertation on construction technology, geology, or hydraulics here, in this courtroom, I would certainly be a very bad technician. I would also be a very bad lawyer.[40]

The position expressed in cassation by the plaintiffs' lawyer Odoardo Ascari was not substantially different, who urged against turning the trial into a sort of university exam, with the defendants called to the blackboard to solve complex equations.[41] In the same vein, prosecutor Arcangelo Mandarino, in his 1967 indictment, made it clear that, for trial purposes, it was not relevant to discuss the scientific impossibility of knowing all aspects of the landslide phenomenon at Vajont. It was sufficient, he argued, to prove that the risk was known—albeit imperfectly—but nonetheless underestimated.[42] To those who insisted on the intrinsic and imponderable dangers of the most varied human activities, Mandarino reiterated that the precautionary principle required "the observance of all norms of prudence compatible with the risk inherent in the dangerousness of the activity."[43] In short, Mandarino asserted, the higher the risk, the greater the prudence and expertise required to mitigate and control it. Lawyer Canestrini was convinced that the Vajont trial was eminently political, bringing an entire production system to the dock: "There are processes that, by their nature, lay bare the foundations of a society and the meaning of an entire historical epoch. . . . The Vajont trial is, by its very nature, first and foremost a discourse on the responsibilities of science, technology, and bureaucracy in protecting the social and political interests of the ruling class."[44]

Reading and rereading it, Canestrini's argument clearly resonates with many ideas that would later become pillars of political ecology and of science and technology studies. The lawyer dismantled the notion of apolitical, unbiased, and pure scientific and technical progress, asserting the inherently political nature of knowledge. Anticipating one of the central tenets of political ecology, he argued that progress is "always in favor of someone and to the detriment of someone else."[45]

Politicized or pure, science was undoubtedly the protagonist in the process. As I have already mentioned, in those courtrooms—as indeed throughout the country—the central question was whether and to what extent the disaster had been foreseen. Since lawyers and criminal culpability were involved, this was a matter not of academic debate but rather of establishing, for example, who had foreseen what and who had knowledge of it.

In a spasmodic attempt to exonerate the defendants, the defense strategies oscillated between predictable blame shifting and metaphysical appeals to the inscrutability of nature and its laws. Although the main argument of the defense was the impossibility of predicting the landslide in both the form in which it occurred and its timing, individual defendants did not hesitate to accuse one another of concealing or underestimating critical data.

This was, for instance, the defense strategy of Professor Ghetti, who was responsible for the Novi experiments described in chapter 2. Ghetti accused SADE of withholding crucial data necessary for the complete and effective design of the experiments—an argument the court upheld, leading to his full acquittal. Similarly, the government officials charged with overseeing the dam also claimed they had been denied access to all the data available to the concessionaire.[46] In particular, the Müller report and the Novi test results were key pieces of information that SADE deliberately chose not to share with all concerned parties. According to the prosecutor, the motive behind these concealments was SADE's desire to maintain absolute control over the decisions to be made regarding Vajont: "It was an inviolable principle and a prerogative of SADE managers to gather information, select it, evaluate it, keep it exclusively to themselves, and decide accordingly. Others, whoever they were . . . were to know only what was sufficient to reassure them. They were to remain unaware of anything that might cause alarm or prompt precautionary measures detrimental to the concessionaire's interests."[47]

The 1969 lower court ruling contradicted the prosecution's thesis, arguing that SADE had publicized, rather than concealed, the Vajont reservoir. In fact, the judges appeared to conflate the public display of the prodigious dam with the disclosure of scientific reports produced by experts.[48] In short, a trip to Vajont could hardly replace geologist Müller's eighteen reports, the conclusions of Edoardo Semenza and Giudici's campaigns, or the results of Novi's experiments.

Although the prosecution emphasized the regime of secrecy—or at least the lack of full transparency—adopted by the company, it nonetheless held other parties co-responsible for the disaster. The prosecution described Violin, the chief engineer of the Belluno Public Works Department (Genio

Civile) as little more than a paper pusher for SADE. Moreover, he was not even an efficient one: On October 8, 1963, the government assistant tasked with closely monitoring the Vajont landslide sent his final and extremely alarming report to the civil engineer. Violin, however, chose to forward it by regular mail to the dam service on the morning of October 9, the very day of the disaster.[49] It is worth noting that Violin was acquitted on appeal due to insufficient evidence, but sometimes responsibility extends beyond the boundaries of the criminal code.

That said, the charge brought against Violin was, in many ways, more broadly applicable to the public officials tasked with overseeing the Vajont Dam. According to the prosecution and much of public opinion, the supervisory bodies had abdicated their responsibilities by relying entirely on the entities they were meant to oversee. Although it is true that SADE withheld certain information, it is equally true that public officials did not appear particularly eager to stay informed.

The plaintiffs' lawyer Carloni highlighted the example of Curzio Batini, president of the fourth section of the Higher Council of Public Works, who, despite being aware of the experiments conducted by Professor Ghetti, never requested the results.[50] Sensidoni is described by engineer Tonini of the SADE research office as an excellent advocate for the hydropower company within the ministries in Rome. He managed to secure authorization for raising the reservoir by fifteen meters in just a few days, between December 21 and 23, 1962.[51] Attorney Carloni described the defendant's behavior as follows: "It is certain that he made prodigious efforts to secure clearance for the reservoirs, including the third one. Equally certain, however, is his complete inactivity regarding the various verification, surveillance, and precautionary tasks that he should have promoted, given his role as the director of the dam service."[52]

The plaintiffs' lawyers and prosecutors had no doubt that the inspectors had relied entirely on the company to monitor the landslide and its potential consequences. Exemplifying this attitude were the actions of the Belluno Public Works Department, the Prefecture of Udine, and the Ministry of Public Works in response to a dramatic letter from the town council of Erto and Casso. The council had requested a clear and definitive answer about the risks posed by the dam.

The response was delegated to ENEL—those accused of endangering public safety—who not only assured the council that the landslide was under control but also mocked its concerns.[53] Prosecutor Arcangelo Mandarino spoke of a "mutual reliance"—in other words, the controllers relied on the controlled—and argued that "the empiricism, recklessness and inexperience of everyone made a decisive contribution to the catastrophe."[54]

For their part, the SADE-ENEL representatives and their defense counsel sought to shift all responsibility onto others—in their case, primarily the consultants who were supposed to guide their decisions regarding the Toc landslide. This was the argument put forth by lawyer Bras, part of Biadene's defense team, who claimed that his client was not responsible for the decisions he made because he had merely followed the advice of the experts.[55] The clash between Biadene and Ghetti was emblematic of this defense strategy. Ghetti accused SADE of failing to provide him with all the necessary data for proper calibration of the model, whereas Biadene defended himself by citing Ghetti's assurances that an altitude of seven hundred meters would remain safe, even in the event of the most catastrophic landslide. However, to be accurate, Biadene had requested a higher reservoir level than the altitude Ghetti considered safe.

When asked by the Parliamentary Commission of Inquiry whether he was aware of geologist Penta's catastrophic scenario,[56] engineer Pancini, the site manager at Vajont, responded—somewhat reasonably:

> But if this memo from Penta shows his concern, why did he allow the encroachment? I keep wondering these days; I lost all my people up there. . . . When this Penta allows us, after seeing all the measurements and everything, he would have made his interpretations. When we asked him for the encroachment permit, why did he allow it? They themselves didn't understand it [the experts], let alone us. If Prof. Dal Piaz had understood it, he would have put us on notice. He never did.[57]

Pancini's torment likely went beyond a defensive strategy aimed at shifting responsibility onto others. In fact, the Vajont site manager would take his own life on November 24, 1968, the day before the trial was set to begin. He would be partially vindicated by the plaintiffs' lawyer Giovanni Carloni. In his plea before the Court of Appeals in L'Aquila, Carloni questioned how "the distinguished geologist" Penta had been able to authorize the reservoir at seven hundred meters, despite his own recommendation not to exceed an elevation of 650 meters due to the possibility—never ruled out—of a catastrophic landslide event.[58] Court document 1757/20 reports a meeting at Vajont on November 15–16, 1960, attended by Pancini, Müller, Semenza, and others, in which the idea of a reservoir higher than 650 meters was ruled out to prevent the landslide from causing "an extremely dangerous wave."[59]

Even more eloquent is the court document listed as number 1999/67. Alongside an accurate drawing of the landslide area, Biadene had handwritten: "flooded no, because otherwise the movement of the landslide would become uncontrollable."[60] Later in the same document, which was evidently one of the key pieces of evidence for the prosecution, Biadene wrote that raising the water level in the basin was dangerous due to the dynamic impact that a fast-moving landslide—this adjective being emphasized by the engineer—would cause.[61] That was in 1960. The 1999/67 document conflicted with a deposition by Biadene from November 1963, in which he insisted on an absolute lack of awareness regarding the extent of the landslide:

> I and my collaborators, when we followed the progress of the phenomenon, always said: here we are facing what Dal Piaz had predicted. . . . In short, we were expecting a series of small landslides. . . . I, along with engineer Semenza, began increasing the water level again, until October 1961. We saw that things were developing in such a way that it did not lead to any accelerations that would suggest the worst-case scenario, but we always expected only small landslides.[62]

In the same deposition, Biadene admitted that after Dal Piaz's death in 1962, SADE no longer had a geologist "explicitly on site." He justified this decision by explaining that the testing commission had required fortnightly reports to be sent to all the overseeing bodies.[63] As I have already recounted, the government assistant sent a very alarmed report to the Public Works Department the day before the disaster—perhaps late, but not so late that evacuation of the settlements couldn't have been considered. However, as I've already mentioned, despite the alarm, the chief civil engineer of Belluno chose to respond by forwarding this report to the dam service—by regular mail. The rest, at this point, is well known.

THE (COMMUNIST) MOLE AND THE PROFESSOR

Another criminal trial took place on the fringes of the Vajont disaster: the trial of Lorenzo Rizzato, a technician and draftsman working at the University of Padua, specifically at the Institute of Hydraulics, led by Professor Ghetti. Rizzato was accused by Ghetti of stealing documents related to Novi's experiments, with the intent of passing them through the Communist deputy Busetto to the press. However, it was not even a case of actual theft, as the documents disappeared on October 12 and were found again, in their place, on October 14.[64] As previously noted, the experiments conducted by Ghetti had a rather secretive, if not outright clandestine, nature. Therefore, it was not assumed that they would be made public, even in the aftermath of the tragedy. The circumstantial evidence that placed Rizzato in the dock was mainly his membership of the Communist Party, which was at the forefront of the battle for the truth about what happened at Vajont. It was a summary trial, and as early as October 22, the judges in Padua acquitted Rizzato due to insufficient evidence. Only in 2001, thirty-eight years after the disaster, did Rizzato admit to stealing the papers related to Novi's experiments to make them public, justifying his actions as an act of civil disobedience.[65]

Particularly odious, then, was the charge on the first page of the criminal case against Rizzato: multiple counts of grand theft "for the purpose of profiting for himself and others."[66] In his confrontation with Benedini, assistant to Professor Ghetti and his main accuser, Rizzato recalled that it was not he, but many other employees at the Institute of Hydraulics, who had mentioned the existence of highly sensitive documents during those days "that could be a source of profit."[67] Moreover, in their acquittal ruling, the Padua judges stated that Professor

Ghetti himself had confirmed, through his statements, the general climate of suspicion that prevailed at the Institute of Hydraulics:

> It is noted—the judges wrote—that Prof. Ghetti says he did not keep the glosses and drawings of the report along with the other Vajont documents locked in his drawer, so as not to arouse 'suspicion or worse.' He does not specify at whom these suspicions were directed, or for what reasons, nor who formulated them and for what purpose. But certainly, an air of suspicion (or worse, academic jealousy) was palpable in the institute surrounding that report.[68]

Despite the fact that Rizzato's accusers were inclined to frame his actions as motivated by his political militancy, the criminal proceedings nonetheless emphasized individual profit. Surprisingly, they culminated in an arrest warrant for the Padua draftsman, justified by "the dangerousness demonstrated by the offender, the risk of escape, and the alarm [*sic*] that could result from the recent national catastrophe." As a result, Lorenzo ended up in jail for a week, where, as his son told me, some inmates affiliated with the Communist Party protected him from potential attacks that were likely politically.[69]

How can one blame journalist Mario Passi, who, writing in *L'Unità*, asked why Rizzato was in jail and not the true perpetrators of the disaster:

> They arrested a young technician from the University of Padua. He is not an executive of SADE, nor is he one of those who concealed the terrifying danger of the Vajont landslide or failed to warn the population. He has nothing to do with these matters. He was arrested on the charge of stealing a copy of a document from the Institute of Hydraulics at the University of Padua—one concerning SADE and the Vajont dam. . . . In the face of the

> tragedy of Longarone, we believe that anyone who can reveal even a shred of truth or offer any small detail that could shed light on what happened must do so. Were these the intentions of the young technician from Padua, who was later released from jail? We don't know.[70]

The purpose of the theft—which more closely resembled a loan than an actual theft, as the documents were returned to their original place—became even clearer when Giuseppe Golinelli, an Italian Communist Party (PCI) deputy, and journalist Mario Passi presented themselves to the Belluno prosecutor with copies of the Ghetti report.[71] The notion of social dangerousness and the motive of profit, mentioned by the prosecutor in Padua, are difficult to see in the actions of those who allegedly took the documents. As expected, Golinelli and Passi claimed they did not know the source of the previously confidential documents.

The trial of Lorenzo Rizzato is, at best, a minor appendage to the Vajont tragedy and a small footnote in its judicial history. The volume of documentation gathered for the trial of those responsible for the massacre is incomparable to the meager case file on the Rizzato matter. As I have already mentioned, the trial, which was held under summary judgment, ended on October 21, 1963, with Rizzato's acquittal due to insufficient evidence. The Padua judges seemed almost reluctant to convict Rizzato, likely acknowledging the noncriminal nature of his actions. In fact, the formula used—insufficient evidence—acknowledged the existence of evidence but found it inadequate for a conviction. The ruling not only discredited the circumstantial evidence—for instance, the supposed "swelling" under Rizzato's jacket to hide the file or his nervousness the morning the report was found back in the institution's drawers—but also addressed the main piece of evidence: his membership of the Communist Party. Once again, for the Padua judges, this was merely a piece of evidence that did not prove the defendant's guilt. The telegrams of solidarity sent by numerous local PCI federations to the detained Rizzato were demonstration of "petty and everyday political polemics" and certainly not proof of guilt.[72] Also discrediting the theorem linking the Communist militant to the theft of the report was the fact that news of the Ghetti report had first appeared in the pages of *Il Giorno* and only later in *L'Unità*—which was the official newspaper of the Communist Party. Moreover, Member of Parliament Busetto, in a parliamentary question on the Rizzato case, referred to a SADE draftsman rather than an employee of the University of Padua. While reading these documents, I was deeply surprised that the judges did not consider the possibility that this could have been a strategy to protect Rizzato. Or, more likely, they chose not to entertain this possibility and instead resorted to acquittal for insufficient evidence, faced with a defendant they did not want to punish.

As I have already explained, in 2001, thirty-eight years after the massacre, Lorenzo Rizzato admitted in public that he had stolen the Ghetti report to hand it over to PCI MP Busetto. It was certainly not out of fear that Rizzato had kept that matter confidential for so long; in four years of trials, Rizzato had been acquitted and ran no criminal risk in revealing his secret. Rather, I believe Rizzato's secrecy was due not to a lack of courage—the Padua-based draftsman had shown he had plenty of it—but rather to an attitude of deep humility. Rizzato did not feel or want to be the hero of Vajont; his had been a minor and at once great gesture of civil disobedience. As always with civil disobedience, it is necessary to force the limits imposed by the law to obey an imperative stronger than the prevailing norm because, as the militant historian Howard Zinn reminds us, democracy always comes from breaking the law, not from blind obedience. Pierantonio Rizzato confirms his father's reluctance to talk about his role in the Vajont disaster.[73] On the occasion of his death in 2019, journalist Fabrizio Zupo reconstructed the story of Lorenzo Rizzato in the pages of Padua's *Il Mattino*.[74] Although acquitted, the trial cost Rizzato a four-year suspension from work—the duration of the various levels of the trial. Four long years in which Lorenzo was left without a salary, forced to "scrape by," as Pierantonio recounts, to make ends meet. At the end of the criminal proceedings, Lorenzo was reinstated at the University of Padua, albeit in a different department from hydraulics. Meanwhile, a new spark had ignited in the designer's life, transforming him into an author, director, and theater organizer. In 1964, Lorenzo founded the *Teatro Popolare di Ricerca* (TPR), a cultural institution of great importance, and not just on a city level or regional scale.

We do not know how the events at Vajont influenced Rizzato's choices, especially his newfound passion for popular research theater. I like to glimpse in this choice the legacy of Brazilian pedagogue Paulo Freire, with his "theater of the oppressed" and his radical stance of humility—his dissatisfaction with a scientific knowledge that does not reveal injustice but rather covers it, and his desire to give voice to the voiceless. Pierantonio confirmed my conjectures, telling me about the influence Bertolt Brecht's work had on his father and his choice of politically aligned theater. Lorenzo dedicated the rest of his life to the TPR, and although he studiously refused any starring role in the Vajont disaster, that story nonetheless stuck with him. Pierantonio, who picked up his father's theatrical legacy, recounts how his father and, consequently, TPR paid a "huge price"—for example, by being excluded from opportunities provided by local government.

If Lorenzo did not feel like the hero of Vajont, even less did he see himself as its victim. With the stubbornness of someone of solid convictions and unshakable values, Lorenzo had simply decided to break the law because, regarding Vajont, he could not remain neutral. That act of civil disobedience changed both the Vajont case and his life.

4

RESISTANCES

REBELLIOUS VICTIMS

When recounting the story of Vajont—or, in truth, any number of other stories from what I have termed the Wasteocene, this era characterized by the production of discarded communities—I have often been accused of promoting a paradigm of victimhood.[1] According to this critique, I portray these communities as victims of injustice and, in doing so, deprive them of their agency. They are seen as objects of others' actions rather than as actors in their own right. Receiving criticism is not always pleasant; at times, it feels harsh, incomprehensible, or biased. In the worst cases, it can evoke petty academic disputes.

However, there is also much valuable criticism—feedback that helps us see ourselves more clearly, revisit our beliefs, and perhaps change them or, at the very least, articulate them more effectively. My reflections on the relationship between the victim paradigm and agency are deeply indebted to the many colleagues who have challenged me to think critically about these issues.

In a poignant 2011 essay, historian Giovanni De Luna wrote about the "republic of grief" and observed how, in recent Italian public history, we have shifted from celebrating heroes to commemorating victims.[2] In a complex and often contradictory rosary of formalized memories, the Italian state has constructed a narrative framework of national identity centered precisely on the memory of victims. As De Luna points out, this is part of a global trend, affecting other countries as well, particularly those with recent histories marked by profound and violent divisions, such as Spain or Germany. The victim paradigm has also been explored by literary scholars such as Giglioli, who has aptly observed: "The victim is the hero of our time."[3] Quoting Giglioli, Andrea Suverato argues that the status of victimhood confers prestige and recognition by establishing a form of absolute innocence.[4]

In many cases, those who critique the victimhood paradigm in these terms are referring to the public use of that narrative device, not to the victims themselves. I think it's equally important to clarify what kind of victims we are discussing. It may seem obvious, but it's worth stating clearly: Not all victims are equal in the marketplace of grief. Not all are privileged with being remembered, nor are all recognized as victims.

Anyone who has studied environmental conflicts, even superficially, knows how difficult it is to prove harm suffered. Establishing a causal link between a polluting source and a disease is akin to the search for the Holy Grail. When you factor in the enormous disparity in power between polluters and victims, the task becomes nearly impossible. This is why I have chosen to focus not on the victimhood paradigm itself but rather on the toxic narrative infrastructure that normalizes, invisibilizes, and naturalizes injustice.

Try visiting the Tamburi district of Taranto, the *Terra dei Fuochi* (land of fires) in Campania, petrochemical Gela, or Marghera and claiming that victims in Italy are granted a hearing, recognition, or prestige.[5] Instead, I agree with De Luna that the victimhood paradigm defuses the conflictual essence of the histories it references. It is easier to remember victims than to identify perpetrators, and even more difficult to address the systemic and structural injustices that create victims in the first place.

The "republic of grief"—to borrow De Luna's apt phrase—may mourn the innocent, but it is far less willing to pursue justice or confront conflict, which are essential foundations for true justice.

Is this not precisely what happened—and continues to happen—at Vajont? I am not convinced that the victims of Vajont received special treatment, either symbolically or materially. Consider the paltry sentence with which the Vajont court case concluded; it hardly seems that the victims were truly heard. Perhaps this merely confirms the grief-centered victim paradigm: Vajont was meant to—and must—signify mourning, not a demand for justice.

That said, in the case of Vajont, the toxic narrative infrastructure has primarily operated through invisibilization and naturalization, given the disaster's minimal presence in the nation's collective memory—at least until the televised airing of Marco Paolini's monologue in 1997.[6] I will address the memory of Vajont more thoroughly in the fifth and final chapter. However, it's clear that the organization of memory—what is remembered and how—is deeply connected to the tension between victimization and agency.

The absurd saga of the two Vajont cemeteries—brilliantly recounted by journalist and writer Lucia Vastano—can be interpreted as a material representation of this friction between the victimization paradigm and a more conflictual approach to the injustice suffered.[7] In 2003, the municipality of Longarone decided to use part of the compensation funds to build a monumental cemetery for the victims of the disaster. The new

Figure 4.1
Cemetery of the Vajont victims (designated a national monument in 2003). Courtesy of the private archive of the Pro Loco Longarone Association, Belluno.

monument, resembling a war memorial with its symmetrical lawn and uniform marble blocks, replaced the old cemetery in Fortogna, which had been tended by the victims' relatives.

In Fortogna, memory was not confined to the formalized structure of a memorial. There, the cemetery preserved the rawness of private grief and the confrontational energy of those who expressed their anger: "Barbarically and vilely slaughtered by human recklessness and greed, they wait in vain for justice for the infamous guilt. Premeditated massacre." Lucia Vastano highlights this plaque from the Paiola family—seven dead, including three children—as an example of the confrontational memory erased by the new Vajont memorial. The new site represents a celebration of the victim paradigm, promoting a composed, pacified grief designed to evoke pity and forgiveness rather than fuel anger or conflict.

Perhaps the words of Carolina Teza, a survivor of the disaster, offer the clearest articulation of the tension between the domestication of victims and their politicization:

> Institutions have done and continue to do everything to divide the good survivors from the bad ones. The good survivors are those who speak of pain, who move their listeners with their stories but know how to stop there. They know when to stay silent and allow the institutions to tell the story, making memory harmless so that it doesn't disturb the economic powers that still prioritize profit over human life. The bad survivors are those who seek justice and fight to ensure that their dead become a reminder to the living—never to forget what humanity is capable of in the defense of profit. The bad survivors are the ones who point the finger at a system that values money over human lives.[8]

The opposition between good and bad victims—to use Carolina's apt expression—emerged immediately after the disaster, manifesting as a dichotomy between mourning and protest, grief and anger. As the Christian Democrats proclaimed in large letters on their posters, the communists were "jackals" exploiting the tragedy of Vajont: In the face of grief, there was no room for political battles.[9] This, to me, exemplifies the victim paradigm, which aligns with my thesis on the toxic narrative infrastructure that silences or normalizes injustice.[10] At Vajont, there were undoubtedly two thousand victims. Yet, for some, this collective immersion in grief sufficed, without raising too many questions about who or what was responsible.

In essence, the victim paradigm strips agency not only from the victims but also, in a way, from the perpetrators. The naturalization and normalization imposed by the toxic narrative infrastructure recast violence as mere accident. It is no coincidence that communists were accused of politicizing the disaster; within the victim narrative, there is no space for politics because there is no room for history understood as a layering of conflict and power.

Yet, at Vajont, everything was political. In this story, which intertwines nature and economy, power and science, technology and memory, the victims are never just victims—they are also rebels.

ON THE BARRICADES

It is not easy to narrate the agency of the victims, the rebellion of the subalterns. One might be inclined to say that it is not just the ill will of mainstream historiography—although, in truth, this does not help. Social history, subaltern studies, gender studies, and postcolonial studies have demonstrated how it is possible to resist the invisibilization of alternative histories imposed by the mainstream organization of memory.

However, one cannot ignore the fact that, generally, these subaltern and marginal histories have left very faint traces. In short, while SADE has preserved a rich archive of its activities, the same cannot be said for the citizens' committees that, at Vajont and elsewhere, organized to defend themselves against the bullying of big corporations or the failures of the authorities. Perhaps these histories are even more invisible when the conflict unfolds in the extreme peripheries of modernity, as in the Vajont mountains.

Citizens[11] of Erto and Casso and the mountain hamlets around the future dam began to mobilize against SADE even before the construction of the reservoir began. The hydroelectric project implied a radical transformation of the socioecological relations that structured the space for both humans and nonhumans in the valley. Even before fearing for their physical survival, the people of the Vajont mountains understood the dam's construction as a threat to their way of life.

Building a giant hydroelectric reservoir meant flooding a vast expanse of land that had been devoted to traditional mountain activities. Although the hydropower narrative portrayed the mountains as a kind of virgin land to be "developed," the reality was quite different: The valleys were already populated with humans, plants, and animals. Turning an alpine valley into a hydroelectric engine also meant dismantling the complex web of socioecological relationships that had shaped the lives of these mountain people and their ecosystem.

Furthermore, the construction of large hydroelectric reservoirs involved not only the flooding of land beneath the reservoirs but also the creation of extensive safety zones along the banks, where access and traditional activities were prohibited. As in any story of capitalist accumulation, this one in the hydroelectric industry began with a massive operation of expropriation and displacement.

For the people of Vajont, SADE presented a stark nonchoice: They could either sell their land to the hydroelectric company or face expropriation. The construction of the Great Vajont was classified as a public utility project, which allowed the company to expropriate land if the owners refused to sell.

The situation was further complicated by the fact that, at Vajont and in many mountain areas, a substantial portion of the land was communally owned. Forests and pastures, which did not fit the logic of private ownership, were managed through a network of community institutions and practices.[12] In this sense, hydrocapitalist accumulation intersected with the typical phenomenon of enclosures—whereas Thomas More described a world in which sheep devoured humans, here it was horsepower that consumed both humans and animals alike. In just a few years, more than half of Erto and Casso's 5,200 hectares of communal land came under the control of SADE.[13]

The mountain people were dispossessed of their collective property by the municipal authorities, who arrogated to themselves the right to sell the lands—pastures and forests—to SADE as if they were the owners, confusing municipal property with communal property. Squeezed between the threat of expropriation for public benefit and pressure from an industrial corporation, the Vajont highlanders were forced to sell their land well below market value: thirty-five to forty liras per square meter for the best pastures, and eighteen to thirty liras for the poorest ones.

The first committee opposed SADE precisely on the issue of land sale or expropriation for the construction of the reservoir. Realizing that it was now impossible to resist the corporation's pressure, the citizens at least tried to secure fair compensation for their property. Led by Paolo Gallo, a doctor and the husband of Erto and Casso's mayor, Caterina Filippin, the committee managed to impose a price of a hundred liras per square meter on SADE for all types of land. Evidently, the committee had more bargaining power than individual mountaineers could muster against a hydroelectric giant, although it should be noted that the price they secured was still lower than market prices, which ranged from fifty to two hundred liras per square meter.

It was SADE's response that confirmed the committee's effectiveness as a tool of self-organization. The hydroelectric company worked to undermine the unity the mountaineers had achieved by offering more advantageous deals to those willing to sell. To further complicate matters, rumors spread that the Filippin family had secretly negotiated with SADE to sell their own land.[14] Although these remain suspicions, it is certain that the mayor eventually decided to abandon the fight against SADE, arguing that when faced with the risk of defeat or a prolonged confrontation, it was better to secure something immediately.

As is well known, grassroots mobilizations often follow a karst pattern—sinking and reemerging periodically, sometimes changing leadership or tactics. Journalist Tina Merlin, perhaps the only available source for reconstructing the history of popular resistance to SADE, recounted how a new committee emerged from the ashes of the first one led by the Filippins. This new committee had a broader and more comprehensive resistance agenda. The struggle was no longer confined to the issue of expropriations but instead extended to the wider impacts of the hydroelectric reservoir on the lives of local communities. The issue was not only the loss of prime land—about three thousand hectares according to the Parliamentary Commission of Inquiry—but also the significant complications to mobility caused by the construction of the reservoir.[15] Tina Merlin vividly described the impact of the dam on the socioecological relations of the Vajont valley:

> Some hamlets of Erto were cut off from the center by the reservoir. They had been connected to the main town by paths that crossed the valley, paths along which farmers moved like squirrels. Many residents of Erto owned land on the opposite slope. How would they manage after the lake was built? They asked for a footbridge to connect the two sides. Initially, SADE promised to build one. Then, leveraging its power and influence, it obtained another concession from the ministry that exempted it from constructing the footbridge. Instead, it proposed a bypass road. For the people of Erto, this meant a long and arduous

journey, especially in winter: for the children from the hamlets who had to travel to school in the main town; for the elderly women going to mass at dawn; and for the farmers who now had to traverse more than three kilometers to work their land.[16]

A 1952 specification mandated the construction of a footbridge to allow the passage of people and animals. However, in 1958, with the increase in the dam's height, SADE abandoned the plan for the footbridge and instead opted to build a road around the lake. Although this was a "scenic road," it was exceedingly long and virtually inaccessible to the mountain residents, most of whom lacked private transportation. The 1958 specifications also required SADE to support the establishment of a public transportation service in the nearby municipalities, but this clause was ultimately ignored.[17] The battle over access to natural resources—pastures, fields, and forests—as well as to schools and other essential services had witnessed the alignment of municipal governments and professional organizations. On August 26, 1956, the municipality of Erto and Casso and the Confederation of Farmers jointly signed a document opposing SADE, emphasizing the need for smooth and rapid communication across the reservoir while criticizing the proposal for a scenic road around it. Their words fell on deaf ears.

Meanwhile, amid lands that had been expropriated or rendered nearly inaccessible, the mountain dwellers began to notice the first signs of danger posed by the dam, often manifesting as cracks in the walls of their homes, as discussed earlier. In response to these growing concerns, on May 3, 1959, a consortium was formed in the dance hall of the Ente Nazionale Assistenza Lavoratori (National Workers' Assistance Organization) by heads of families in the municipality of Erto and Casso. Its purpose was to defend the population against

SADE's oppressive actions concerning expropriation and to address potential harm to property and lives from landslides and subsidence along the lake's banks.[18] We know little about the activities of this consortium, which most likely aimed primarily to secure institutional support for its protests against SADE. The Parliamentary Commission of Inquiry mentioned "letters of complaint, discussions within the collegiate bodies of local authorities" and parliamentary interventions.[19] It would seem, then, that the protests at Vajont were progressing along the established paths of electoral representation and law. However, commenting on the founding assembly of the consortium, Tina Merlin wrote: "The assembly was held in an orderly fashion, but in an atmosphere of rebellion that everyone had been harboring in their hearts for some time. An old woman said, 'If thieves come to steal in my house, I have every right to take my rifle and defend myself.'"[20]

The tense atmosphere at Vajont was best confirmed by the disproportionate and repressive reactions of the security authorities aligned to defend the corporation's interests. Tina Merlin recounted, for instance, how the *carabinieri*[21] attempted to intimidate Giovanni Martinelli, who attended the consortium's assembly carrying two signs: "Down with SADE" and "Down with the government." For the police, apparently, these messages were unacceptable. Martinelli, whose house had been destroyed by the Nazis during the war for aiding partisans, was not someone easily intimidated.

The real question, however, is what the *carabinieri* were doing in a remote alpine valley. In 1956, three years earlier, a *carabinieri* post had been established in Erto and Casso. Given the absence of a surge in criminal activity in the region, it is reasonable to infer that their presence was primarily linked to the construction of the dam and the protests surrounding it. Journalist Mario Passi, for example, recounted an incident where the *carabinieri* detained Belluno PCI member Giorgio Bettiol, a member of the consortium, and brought him to the station, warning him not to spread "false" information among the mountain people.[22] It was likely the same *carabinieri* from Erto and Casso who filed the complaint against Tina Merlin and *L'Unità*, as I discussed in detail in chapter 3.

The majority report of the Parliamentary Commission of Inquiry claimed that the mobilizations before the disaster primarily centered on defending the economic interests of the local communities. This interpretation aligned with the broader narrative of the report, which sought to demonstrate that the disaster was unforeseeable. Yet, despite its evident reticence, even this report was forced to acknowledge that the grievances voiced by the people of Vajont often referenced the "grave danger to human life." The commissioners, however, took great care to emphasize that the dangers perceived by the local population were "incorrect"—that is, not aligned with the actual events that would later unfold.[23]

As I mentioned earlier in this book, the accusation of geological inaccuracy leveled against those who voiced concerns about the risks posed by the dam is particularly striking, as if it were not the responsibility of the authorities to conduct the necessary checks with the expertise and resources at their disposal. Whether accurate or not, the fears surrounding the massive dam were a significant motivation for the mountain people's resistance against SADE. Both Tina Merlin and the parliamentary committee, for example, acknowledged the role of the Pontesei landslide in November 1960 as a catalyst for a new phase of mobilizations focused on safety concerns. Journalist Renato Zanolli wrote about a popular oversight committee formed in the aftermath of the Pontesei landslide.[24]

However, once again, we are left with only faint traces of this grassroots activism, which contradicts the narrative of the mountain people's acquiescence. Renato Trevisol, a SADE technician, recalled that shepherds from Erto would often go to the dam to protest: the microshocks and rumbles that shook the Toc frightened their animals, causing them to refuse to go to pasture.[25] In the same interview, Trevisol recounted the horror of witnessing bodies mixed with mud in the aftermath of the disaster. These were victims, no doubt, but also rebels, because those lifeless bodies belonged to the very people who had protested, rejecting the normalization of fear and risk imposed by the corporation.

JUSTICE BEYOND THE APOCALYPSE

For those who survived the flood, the challenge of demanding justice for their loved ones killed by the disaster remained. This battle was waged not only in court, as discussed in chapter 3, but also in parliament, in the streets, and through the newspapers. The post-disaster period brought with it a number of concrete issues that demanded mobilization, such as the reconstruction and relocation of displaced residents and the fight for fair compensation. What happened to Vajont after the disaster has been the subject of numerous studies and impassioned reconstructions, detailing legislation, corruption, the rebuilding process, and the psychological and social consequences of the disaster on survivors.[26] I would like to remain within the framework of this chapter and focus on the various forms of resistance and mobilization that emerged in the aftermath of the disaster. The most striking post-Vajont protest—the one that has left a trace in the written memory of the tragedy—was the demonstration in December 1963. The first survivors' committee organized a roadblock near Dogna, preventing tourists from reaching Cortina for the Christmas holidays.[27] Evidently, only a few months after the massacre, a bold and dramatic gesture was already necessary to bring Vajont back into the national spotlight. The documentary *Vajont: 2000 Condemnations*, produced by the PCI Film Office in 1970, included archival footage of the December 31 roadblock.[28] Aside from the evident tensions with the police—signaling a waning solidarity with the survivors—the voices of those involved in the protests emerge vividly in the PCI documentary. One woman, interviewed at the protest site, recalled the widespread awareness of the Toc rock's instability and declared: "We want the dam to be removed as soon as possible; otherwise, we will take it down ourselves."[29]

Another man recounted how, two months after the disaster, survivors were left with no prospects for the future, leaving them no choice but to stand on the barricades.

Many of the protests centered on the issue of relocating settlements, with survivors striving to defend the possibility of rebuilding their flood-ravaged villages locally. In a detailed interview with journalist Marina Zerman, Elsa Capraro, a Longarone switchboard operator who survived the disaster, recounted the roadblocks organized during the winter of 1963–1964 to resist the final abandonment of the villages destroyed by the flood.[30] A January 1964 newsreel also covered the protests, capturing signs held by demonstrators that condemned the bureaucracy for "killing Longarone a second time" after the flood. The signs also vowed to honor the dead by refusing to allow the town to be relocated elsewhere.[31] Still the documentary produced by the PCI showed footage of a march held on November 24, 1963, with the participation of other communities threatened by the presence of hydroelectric reservoirs.[32] The footage depicted a massive demonstration, unified by a central slogan displayed in various forms on the placards: "Justice for the victims." Alongside the defense of their roots, the demand for justice remained a cornerstone of the post-disaster mobilizations. Journalist Fiorello Zangrando recalls a demonstration held on January 10, 1970, which concluded at the Belluno Municipal Theater. The protest was convened to prevent the criminal trial from being nullified by the statute of limitations for the defendants.[33]

Building on Tina Merlin's work and direct involvement in the events leading up to the disaster, the PCI continued to play a leading role in demanding justice after October 9, 1963. The case of Lorenzo Rizzato, the Padua draftsman introduced in chapter 3, further demonstrated the PCI's commitment to uncovering the truth. On October 13, 1963—just four days after the tragedy—the Communist Party presented a White Paper on Vajont to President Antonio Segni. This document compiled evidence highlighting the foreseeability of the catastrophe. From the very first page, the pamphlet clearly articulated the demand for justice: "Honorable President, with the sole humane and civic intent that justice be done and that all those responsible for the Vajont tragedy be held accountable, we have gathered initial documentation that testifies to and proves that the disaster could and should have been avoided."[34]

The White Paper brought together a wide range of statements, primarily from PCI members but also from local authorities, regarding the dangers posed by the dam. Although it is true that some documents focused more heavily on economic concerns—seeking to protect the livelihoods of the mountain communities rather than explicitly addressing their survival—there were also numerous interventions that emphasized the issue of risk. For instance, the document includes a proposal from February 1961 by a group of provincial councilors from the PCI and the Italian Socialist Party. They suggested commissioning Professor Gortani, a Christian Democratic senator and renowned geologist, "to conduct a thorough investigation of the Vajont situation."[35] The entire provincial council aligned with the minority, endorsing a resolution that called for implementing all necessary safety measures to guarantee the well-being of Vajont's citizens.[36] The White Paper, of course, included the aforementioned letter from the mayor of Erto and Casso, dated September 2, 1963, addressed to all stakeholders, in which the local administration unequivocally expressed the population's fears. ENEL-SADE responded to this letter in a dismissive and reassuring tone.[37]

In city and provincial councils, in parliament, and in the pages of their media outlets, the Communists launched a campaign to identify those responsible for the disaster. As the party's Central Committee declared: "To honor the victims of Vajont, to show solidarity with the survivors, means, above all, to adopt the urgent popular demand to uncover the criminal motives of monopolistic profit and government complicity that lie at the root of the tragedy, and to ensure that justice is inexorably served against the guilty, up to the highest levels of both technical-administrative and political responsibility."[38]

For the mainstream press, the Christian Democrats and their allies, the PCI was accused of instrumentalizing the pain of the survivors. By demanding justice, the Communists were politicizing Vajont; in the wake of the disaster, the focus should have been on collective mourning for the victims, not on dividing people by pointing to the responsibilities of the corporation and the government. Mourning, in their view, involved neither anger nor justice. Clearly, Vajont was political beyond the struggles of the PCI because there was nothing apolitical or natural about what had occurred in that valley. In essence, the communists were exposing the layers of power that had shaped the landscape before and after the disaster. We will later examine the implications of the PCI's stance in the search for an official truth, particularly the party's decision to submit a minority report to the Parliamentary Commission of Inquiry. It is more challenging to recount the grassroots mobilization that also accompanied the PCI's efforts at Vajont. Hostile sources, however, offer some insight; many journalists accused the communist militants of invading Vajont from nearby "red" regions.[39] For instance, Alberto Cavallari described the arrival of communist militants from Emilia Romagna in this manner.[40] Indeed, many volunteers, often former partisans, arrived at Vajont to offer aid and solidarity. This was not a temporary commitment but a lasting one; for instance, Canestrini, the principal lawyer of the plaintiffs, was closely tied to the PCI. One story I would like to tell, if only I had the sources, is that of the survivors' trips from Vajont to L'Aquila when the trial was moved to the Abruzzo capital due to legitimate suspicion. The journey became an additional hardship imposed on the survivors, who were forced to choose between enduring the expense and inconvenience of travel or forfeiting their chance to attend the trial. Sparse and fragmentary news, however, provides us with a picture of people who were not merely passive victims but rather actively engaged in the pursuit of justice. The journey turned into an opportunity to build solidarity with the CGIL (Italian General Confederation of Labor), which seems to have provided the buses, and with other communities along the route that welcomed the Vajont survivors. Although we can only imagine what transpired during those trips—who made them possible, who organized and participated in the meetings—it seems plausible to assume that major political and labor organizations, which had supported the Vajont cause from the left, played a significant role.

The final act that solidified the PCI's position in the Vajont struggle was its refusal to sign the majority report of the Parliamentary Commission of Inquiry. As early as October 19, 1963, Senator Scoccimarro delivered a lengthy speech in parliament on Vajont, in which he outlined the principles guiding the PCI's stance. Against the rhetoric of resignation, which had been invoked repeatedly in the name of appeasement and mourning, Scoccimarro reminded the assembly that, in the face of the disaster's predictability, "offended humanity does not resign, but rebels; and thus, you have the revolt and the eruption of popular sentiment demanding justice."[41]

While insisting on the need to ascertain the specific responsibilities of public and private officials, Scoccimarro proposed moving beyond an entirely technical view of what had happened: "To answer the dramatic question that confronts us today, we must . . . go beyond the technical bounds where, Minister, you seem to want to stop; in short, we must look deeper into the political substratum of this painful affair and identify its original cause. And that cause lies in the logic and spirit of the great capitalist monopolies."[42] This reflection on technology—or rather, techno-fixes—extends beyond the Vajont Dam, Longarone, SADE, and even the Italy of the economic miracle and the Cold War. Instead, it speaks to the contemporary socioecological crisis by unmasking not only the illusion of techno-fixes—the belief that technological innovations can solve the crisis—but also the instrumental use of a crisis narrative focused on the responsibilities of technology or, perhaps, the guidance of science. Then, as now, the issue is not about dispensing with expert opinions or technical verification but rather about ensuring that the space for politics is not nullified. The idea that technology or science can replace politics by imposing a neutral, conflict-free agenda is either an illusion or a fiction. In the case of Vajont, as Scoccimarro noted, expert knowledge revealed conflicting opinions about the stability of Monte Toc. However, this conclusion left open equally divergent interpretations. The government and its majority believed this exonerated them from responsibility, whereas the PCI, through Scoccimarro's words, questioned, "Indeed, when the technical investigation revealed that there were originally conflicting opinions among the technicians, was this uncertainty not reason enough to avoid the terrible risk of such a major calamity? Instead, no, they proceeded anyway."[43]

According to the Communist senator, the disregard for any precautionary principle was intrinsic to the nature of a certain type of capitalist enterprise, one that pursued profit with a "spirit of adventure," as he put it. This argument was later taken up and further developed in the minority report submitted by the PCI to the Parliamentary Commission of Inquiry. Communist representatives rejected "the appeal to fatalism or the unleashed forces of nature." They reiterated the need for political judgment regarding the lack of precaution and the recklessness displayed by SADE and the regulatory bodies.[44] The criticism of a purely technocratic approach became not only an indictment of the hydroelectric monopoly and the subordination of political power but also a proposal for a new organization of the relationship between technology, power, and politics. In its minority report, the PCI emphasized the need to democratize and decentralize decisions regarding land use, prioritizing community needs over economic development goals. The call for the sovereignty of elected assemblies and broader democratic participation did not, however, entail a rejection of expert knowledge. On the contrary, the Communists advocated for strengthening public oversight by hiring experts who could guarantee both independence and competence.[45]

The PCI has left a substantial record of its involvement in the Vajont disaster; however, it is more challenging to trace the involvement of other social and political actors, particularly those less structured than a large mass party. For example, the attitude of the extra-parliamentary left regarding the Vajont disaster still needs further exploration. The online archives of some radical leftist newspapers do not seem to show much interest in the disaster, but obviously, more in-depth research using various sources, especially oral accounts, would be required.

The conflicts and the resistance at Vajont, however, did not end with the L'Aquila trial and the legal determination of responsibility for the disaster. In more recent years, Vajont has been marked by conflicts focused primarily on the preservation of the disaster's memory and the search for a narrative or restorative justice that extends beyond courtrooms, financial reparations, and legal punishments. In short, these are conflicts over shared and conflicting memories, which the next chapter will address.

5

MEMORIES, SHARED AND DIVIDED

PRESERVING MEMORY / RECLAIMING MEMORY

I have often heard, sometimes in a patronizing tone, that history can be useful. It is repeated that we need to remember the past to avoid repeating its mistakes or that memory must be cherished and passed on to future generations. This becomes especially important when it comes to remembering past disasters because then memory is expected to serve as a kind of instruction booklet. Rather than helping us understand how to construct the present, we are warned about how it might go wrong if we are not careful. Perhaps it is more like a gallery of horrors than an instruction booklet.

Although I am convinced that history is useful—and not only because it provides me with a living—I am not as convinced that its usefulness lies in this "freezer function," as if our task as historians were merely to preserve memories, frozen, neatly packaged, and ready for consumption. Instead, I believe that history can serve to denaturalize the present, to free us from a grim realism that turns it into an inescapable necessity. Remembering serves a purpose—those who seek to erase uncomfortable histories of oppression and rebellion know this well—but it is not a simple exercise in memory. Putting something under a glass case or, more prosaically, in the freezer at the department store of our history is not enough.

Memory is never singular, fixed, or definitive; it often refers back to different meanings and values, even when the facts being remembered are the same. Rather than freezing memories, historians can decipher them, training us to recognize the "memorial regimes" that organize what should be remembered, what should be forgotten, and who has the right to tell these stories. A memorial regime establishes the official narrative, which, more often than not, serves to invisibilize and naturalize injustice and violence.

Memorial regimes are particularly relevant when we talk about disasters. The invisibilization and naturalization of injustice and violence are basic ingredients in the recipe for a perfectly depoliticized memory of disasters. Often, this mainstream narrative is the one preserved through public ceremonies, monuments, and all the paraphernalia that accompanies a narrative regime. That is why the task cannot be to "preserve memory" in the singular but rather to challenge this mainstream memory with a multiplicity of insurgent, irregular memories.

It is precisely in the spaces where conflicting memories collide that meaningful insights emerge. Analyzing the making and unmaking of different memorial regimes and the ongoing attempts to sabotage them reveals history's social function.

In the case of disasters, historical series can help plan the construction of infrastructure and the design of urban space. In theory, one should not build a bridge in an area of high seismic risk or a neighborhood in what was once a streambed. Reasoning about memorial regimes, though, goes beyond the statistical reconnaissance of past events. It involves analyzing the ways in which disasters are remembered or forgotten and the conflicting narratives that produce memories and meanings.

As environmental humanities scholar Serenella Iovino aptly observes: "In the face of disasters we have a duty to arm ourselves in a relentless resistance, put in place a moral and cultural militancy so that the fragility they reveal to us is recognized."[1]

In the collective memory, the Vajont disaster has largely been overlooked; it is absent from textbooks and even from some comprehensive works on recent Italian history.[2] For example, there is no trace of Vajont in Guido Crainz's nonetheless fundamental history of the Italian economic miracle[3], although the same author dedicates some poignant pages to the disaster in another volume.[4] Neither Paul Ginsborg's otherwise comprehensive history of contemporary Italy includes a reference to the disaster.[5] Although devoted precisely to disasters in the history of Italy, the volume edited by John Dickie, John Foot, and Frank Snowden does not include a chapter on Vajont.[6] It seems, indeed, that the case of Vajont—with two thousand lives sacrificed on the altar of the country's modernization—has been denied a place in the collective memory; a story too unmodern, almost a critique of the modernist project itself, written in concrete, rock, and mud.

What would it mean to cultivate—not merely preserve—the memories of Vajont? Which memories are granted legitimacy and which are suppressed? What do we truly remember when we commemorate Vajont? How can we disrupt a memorial regime that first erased and then domesticated Vajont's memories and instead pursue narrative justice that restores space to the suppressed memories of the subaltern?[7] When I speak of "narrative justice," I am certainly referring not only to the issue of making the Vajont disaster visible in our collective memory but also to the many attempts to obtain symbolic compensation for the injustice suffered—for example, through the periodic requests for apologies made by the survivors to the Italian state or ENEL.[8]

The inclusion of the Vajont trial archive in the UNESCO Memory of the World project in May 2023 marks a crucial step in the construction of a collective memory of the disaster.[9] And it is, in my opinion, significant that it is precisely a trial archive, emphasizing the contentious nature of memories that have never been reconciled and are still seeking justice.

I am convinced that an entire book would be necessary to explore the construction of both oblivion and remembrance surrounding the Vajont disaster. However, in the following pages, I simply present a triptych that I hope will encourage a rethinking of the relationship between Vajont memories and the subaltern communities' demand for narrative justice.

THE KNOT IN THE HANDKERCHIEF

With these words, Marco Paolini introduced his and Gabriele Vacis's monologue on Vajont, broadcast by Italian public television on October 9, 1997:

> I don't know why you tuned in tonight to hear this story. In some newspapers, this performance is referred to as a documentary; in others, as a drama. Instead, you are in a theater. These people [Paolini is indicating the audience in front of him] are here to listen to theater. For one night, we are stealing the scene from current events and doing a live broadcast about memory. That's why it's so hard to write it, to define what we're doing tonight. Maybe there is no such thing as a live broadcast on memory. What is it? Well, in short, it is a kind of knot in the handkerchief that we have already tied. . . . The story of the Vajont Dam, which began seven years earlier, ends in a few minutes of apocalypse with the holocaust of 2,000 victims.[10]

Marco Paolini and Gabriele Vacis's civil oration marked a turning point in the construction of the memories of Vajont. With more than three million viewers, Paolini's show broke through the oblivion in which Vajont had been buried. As a historian, I felt—and still feel—a sense of shame because it fell to an actor to do what we as scholars should have done or at least what we could have helped to accomplish. As Paolini declares, his civil oration is a "live broadcast about memory," a knot in the kerchief to ensure we do not forget what happened at Vajont.

One could argue that Paolini succeeded in the oft-invoked task of preserving memory, but that would be a mistake. Even though Vajont had been forgotten, sidelined in the grand narrative of Italy's economic miracle, remembering it is never straightforward. It does not refer back to all of the same memories or arguments. In a crucial passage of the monologue, explaining the sources he used to reconstruct the Vajont story, Paolini states: "I am tendentious; you know very well where I stand. It's not like I'm trying to hide my point of view."[11]

This echoes the accusation of tendentiousness that brought Tina Merlin to court in Milan, as we explored in chapter 3. Paolini's civil oration constructs a biased memory. In this sense, the decision to transform the Vajont Dam into a theater of memory—the performance was held outdoors in front of the dam—was not merely a theatrical choice. It embodied the author's stance. Paolini did not aim for a neutral and compassionate memory of a national tragedy but instead constructed a situated, partisan memory aligned with the victims, those who resisted, and those who were ignored in both life and death.

It is no coincidence that Paolini begins and ends his monologue by addressing the mainstream narratives of the disaster.

He recounts the attempts to construct a pacified memory of Vajont. In the aftermath of the tragedy, prominent figures in Italian journalism, such as Dino Buzzati and Giorgio Bocca, described the event as an unpredictable natural disaster, effectively absolving public authorities and SADE of responsibility. In the pages of *Corriere della Sera*, Indro Montanelli[12] accused the communists of being jackals attempting to profit from a terrible national grief.[13] In the final lines of his monologue, Paolini imagines these famous journalists silencing Tina Merlin as she futilely tries to highlight the predictability of the disaster: "Mrs. Merlin, there are the dead; this is not the time for political speculation."

By opening and closing with a reminder of this toxic narrative infrastructure—one that naturalized and erased the injustice of Vajont—Paolini's civil oration is unmistakably positioned in an antagonistic, counter-hegemonic dimension. It is therefore a matter not of preserving memory in a vacuum, lest it be lost, but rather of mobilizing the rebellious memories that were silenced and deserve recognition precisely because they challenge the domesticated mainstream memory—so sanitized that it comes to closely resemble oblivion.

Paolini's narrative undoubtedly stirs pity, particularly in the final moments when he vividly describes the apocalyptic destruction of Longarone. Yet, I believe pity is neither the central register of the monologue nor its ultimate goal. For me, the Vajont oration has always provoked a profound sense of anger. This anger arises, for example, when Paolini labels the ENEL-SADE technicians who insisted on exceeding the seven-hundred-meter safety altitude recommended by Professor Ghetti as "bastards," or when he passionately screams from the stage, demanding to know why Longarone was not evacuated.

For someone in my field, Paolini's indictment of the Italian academy's complicity with SADE can only evoke both anger and enduring shame. Suspended between pity and anger, the entire monologue reflects the complexity I discussed in chapter 4 concerning rebellious victims. Paolini, too, calls for a restoration of the Vajont villagers' agency. He recounts their resistance to expropriation and envisions the formation of the consortium for the rebirth of the Erto Valley as a moment of triumph. Even the apocalyptic events of October 9, 1963, did not annihilate the resistance of these communities. Recounting the dismemberment of the municipality of Erto and Casso after the disaster, Paolini reminds us that if a community remained there, clinging to these mountains, it was because "a group of stubborn individuals . . . enters the valley, steals electricity from ENEL, occupies the old houses, and claims the right to start living again there, where everything had been interrupted."[14]

If the Vajont mountaineers were not only victims but also rebels, then the memory of Vajont should evoke both pity and anger. This is not about mourning the victims of an unfortunate natural accident, as the mainstream narrative would have us believe. Yet, perhaps it is also not about mourning the consequences of a thoughtless criminal act. The rebellious memory evoked by Paolini on the Vajont stage challenges the very foundations of narratives about progress and modernity.

This begins with his irony about the typical school textbook, which always portrays nature through an extractive lens, and extends to his critique of male-dominated enthusiasm for technological gigantism—the belief that the secret of progress lies in a dam or an aircraft carrier. Paolini reveals that the contradiction of Vajont is not just about the disaster itself but also about a mode of production that prioritizes profit above all else.

Perhaps Edoardo Semenza was correct when he criticized Paolini for aligning with Tina Merlin, claiming that both approached Vajont through their own political beliefs. For the geologist Semenza, this politicization distorted their conclusions. For me, however, this was the only way to distort, sabotage, and disrupt the mainstream narrative about Vajont. Rebellious memories can only be partisan.

ON THE BIG SCREEN

In 2001, the story of Vajont made its way to the big screen in a film directed by Renzo Martinelli. This major production featured a stellar cast, including actors well known to the Italian public, such as Laura Morante, Leo Gullotta, Michel Serrault, Daniel Auteuil, and Philippe Leroy.[15] With a budget of eighteen billion lire (approximately €13.7 million today when adjusting for inflation), the film was a significant cinematic undertaking. As with Paolini's monologue, a highly symbolic date and location were chosen for the premiere of Martinelli's film: It was screened on a giant screen near the dam on October 10, 2001, with the audience seated on steel stands set up on the landslide. Once again, the decision to begin at the dam clearly marked the positionality of the film director: Martinelli, like Paolini, aligned himself with the mountain people by supporting the thesis of the disaster's predictability. Omens and signs of impending catastrophe permeate the entire cinematic narrative; one could even say these omens form the thread that unravels the story. Earthquakes, wide fissures in the pastures, cracks in houses, and symbolic forewarnings, such as the wooden statue of a crucifix floating on the reservoir's waters, all underscore the narrative.

The trial of Tina Merlin in Milan is a pivotal moment through which Martinelli highlights how well known the dangers posed by the dam were to the valley's inhabitants. Tina Merlin is undeniably central to Martinelli's reconstruction; her articles become the focal point of discussions in the inns, while everyone—from the parish priest to SADE officials—seems to carry a copy of *L'Unità* in their pockets. Perhaps in the interest of balance, Martinelli includes the figure of the parish priest in his narrative. Initially attempting to engage with SADE, the priest later urges the people of Erto to revolt, offering what could be interpreted as preemptive absolution: "To rebel against SADE is not a sin."

Although it is true that Merlin herself had reported the parish priest's support for the establishment of the consortium defending the Erto valley, the film's depiction of a handshake between the communist journalist and the parish priest feels more like a symbolic gesture. It seems to blend the Don Camillo and Peppone dynamic with the historical compromise (*compromesso storico*) that was yet to come.[16]

If Tina Merlin is the unquestioned heroine of Vajont, courageous in standing up against one of the most powerful corporations of the time, Martinelli attempts to complicate the opposing side by differentiating between the various parties involved in the incident. To simplify things, we could say that Biadene embodies the villain—an unpleasant character in all his manifestations, such as when he takes over engineer Semenza's office after the latter's death. Martinelli portrays him as a man without scruples or doubts, always at the forefront of the silencing of any opinion that might challenge the work and interests of SADE—that is, at least until the final minutes of the story, when we see a doubtful Biadene, assailed by uncertainty, seeking reassurance in any form he can get as he sits in prayer in a church or frantically searches for geologist Penta by phone.

In the cinematic narrative, Pancini—the site foreman at Vajont—becomes almost an anti-Biadene, portrayed as a mild-mannered, emotional man, close to the workers at the site and, above all, consumed by doubts about the dam and the Toc's resilience. Although Pancini also rushes to reassure surveyor Montaner that there is no danger to Longarone, it seems to me that the film invites a more sympathetic judgment of him. Perhaps this reading of Pancini's character is linked to his tragic decision to commit suicide before the trial began. This is a gesture that Martinelli seems to interpret as the culmination of a moral struggle that would have accompanied the site manager throughout the affair. This is, of course, conjecture, but it fits well into the film's plot alongside the sinister portrayal of Pancini's counterpart, Biadene.

In between the two is the figure of Carlo Semenza, who is more difficult to decipher in Martinelli's portrayal: certainly not a "good guy" in the story but not a villain par excellence either. Whereas Biadene insists on the economic interests at stake, Semenza seems more focused on the success of his grand engineering project and the challenge of configuring a balance between technique and nature. Although he prefers to trust Dal Piaz's optimistic views over those of his son Edoardo, Semenza nonetheless seems overwhelmed by anxiety about what is happening at Vajont to the point of death.

Crucial to Martinelli's narrative is the figure of engineer Desidera from the Belluno Public Works Department. The character appears only briefly in the film, as indeed his role in the Vajont affair was minor. Yet, I believe that Desidera occupies an important position in Martinelli's narrative. He is proof that the established order—the institutions and mechanisms of a capitalist society—had the potential to prevent events such as Vajont, if only people such as Desidera, Merlin, the parish priest, Edoardo Semenza, and perhaps even Pancini had been listened to. In some sequences, especially in the phone calls between Venice and Rome, the tone of the narrative shifts toward denouncing connivance and corruption, although never explicitly. Perhaps this is the fundamental difference between Paolini's Vajont and Martinelli's. The former reads Vajont not as an aberration of the system but rather as a manifestation of how the system works, prioritizing profits over the lives of subalterns. Martinelli, on the other hand, seems more inclined to point to individual responsibilities.

The reception of Martinelli's film has been more controversial than Paolini's monologue. Some have pointed out historical errors in the filmic transposition of the story, whereas others have criticized the inclusion of a romantic subplot. Martinelli's film is not a documentary, and the differences between historical reality and the film's narrative choices, I believe, are not errors but rather deliberate decisions. I do not think they alter the historical facts in any radical way. The director is correct when he revisits the internal conflicts of the communities, divided between progress—represented by the dam and wage labor—and the fear of losing everything, literally wiped out by modernity.

Watching the film, at times, a sense of appeasement prevails, with very few truly negative characters—certainly Biadene, perhaps the geologist Penta—and a potential alliance of the "good guys," from Merlin to the parish priest, via Desidera and Pancini, which, however, remains unrealized. However, the last frames of the film reaffirm a sharper stance, or a tendentious one, as Paolini would say. In a lunar Longarone, reduced to an expanse of stones and mud, surveyor Montaner, who had trusted progress by working on the dam, recognizes his house, completely destroyed, by the remnants of the rocking chair given to him by Pancini as a wedding gift. Gripped by grief over the death of his wife and the child she was carrying, Montaner destroys the rocking chair in a scene that perhaps—and at least I like to interpret it this way—undermines the somewhat naive alliance that had been present as an undercurrent in the story.

The film opens and closes at the Vajont cemetery with a line from Montaner at the grave of his wife and unborn child: "Occasionally someone asks me if I have forgiven them. I haven't. I have not forgiven. I can never forgive the men who allowed this to happen."[17]

In the end, Martinelli's proposed narrative is less pacified than one might think.

SIGNS IN THE LANDSCAPE

It is no accident that Martinelli's film opens and closes with the Vajont cemetery. Nothing better symbolizes the shared and divided memories of a great catastrophe than a cemetery. Individual pain flows into a collective suffering that materializes in a citadel of memory. Not everyone has the privilege of a cemetery in which to mourn their dead and establish a bridge between memory and the present. Many Indigenous peoples who have paid the price of modernity and colonialism, not only by losing their lives but often by being dispossessed of their ancestors' sacred burial grounds, know this well. The migrants who survive shipwrecks and do not even have anywhere to mourn their drowned loved ones at the borders of Fortress Europe also know this. Even at Vajont, burying the dead was a privilege: Only about half of the victims were found and buried. The rest of the cemetery is an optical illusion, headstones and crosses without bodies, disembodied memories without even a shred of matter to attach to.

Perhaps it was precisely this absence of the materiality of so many bodies that had filled, almost by counterpoise, the Fortogna cemetery with a dense, palpable memory that manifested itself in the headstones and in the care of the survivors. The case of the Vajont cemetery—which I have already evoked in chapter 4 and writer Lucia Vastano has brilliantly uncovered—seems almost tailor-made to exemplify my argument about disaster memories, and the limits of mainstream discourse in preserving a supposedly neutral but, in fact, divided memory. The Vajont memorial, like all memorials intended to produce a nationalized and pacified memory, is constructed as a neutral place where grief must be composed and anger anesthetized. The monumental inscription that welcomes visitors to the memorial itself seems poor, almost not up to the task assigned to it:

> First the roar of the wave
> Then the silence of death
> Never the oblivion of memory.

Yet, the shared and divided memories are not easily tamed, and like everything at Vajont, the destroyed cemeteries and tombstones also become a matter of conflict, appearing once again in a courtroom—perhaps under the illusion that a judge can respond to a desire for narrative justice that no law will ever be able to ensure.[18]

When reasoning about the Vajont memorial, what is present is as striking as what is absent. Missing, for example, is a monument to the victims of the disaster—not that they are completely absent, but they do not seem to me to deeply mark the landscape. In the monumental cemetery, there is a sculpture by artist and writer Mauro Corona dedicated to the survivors. In the town of Vajont, a large fountain, evoking the shapes of the lake and the dam, has the names of the victims engraved upon it. There is no statue dedicated to Tina Merlin, although streets, squares, and classrooms have been named in honor of the Vajont protagonist. In 2021, Federico D'Incà, minister for relations with Parliament, proposed dedicating a monument to the journalist in his hometown of Belluno, but the proposal was not pursued.[19] About this, we should probably not be surprised, considering the great lack of prominence given to female figures in the organization of Italian collective memory. Even more important, then, is the role of the Tina Merlin Cultural Association, which has been keeping the journalist's memory alive since 1992.

Monuments, toponyms, cemeteries—the museum dedicated to the disaster cannot be absent from this factory of official memory building. Opened in 2009, the Longarone Vajont Museum *Attimi di Storia* was created through the cooperation of the municipality, the Vajont Association "The Future of Memory," and the local cultural/tourist association Pro Loco. For this section, I am indebted to Giacomo Braulin, a student of mine in the graduate program in environmental humanities at Roma Tre University, and his meticulous research on Vajont memories.[20] The museum offers visitors an almost teleological account of Longarone and the Vajont valley, framing it as a tragic preparation for the apocalypse of 1963. On the other hand, the museum's very purpose is to remember that exceptional event, and this mission conditions the way the memories of the place are organized.

If the valley's subaltern relationship to Venice and its political and economic power is a long-standing factor, the arrival of SADE is represented as a watershed moment in the history of the valley and Longarone. The second section of the museum, dedicated to the post-SADE era, is dominated by technical and scientific data on the dam and the landslide, although the figure of Tina Merlin emerges almost disruptively, cutting through the expert knowledge and offering a prophetic glimpse of what is to come. To evoke the memory of the disaster, the museum mobilizes a multisensory experience: Visitors enter a narrow corridor where a disturbing noise guides them toward their rendezvous with the apocalypse. In the final section, the museum recounts the aftermath of the disaster, the trial outcomes, and the commemorations, including references to works by Paolini and Martinelli.

According to Braulin, the museum remains captive to some traditional models of memorializing and mourning, as well as to a romanticized view of pre-SADE socioecological relations. Viewed from a distance, however, the museum seems an interesting experiment in responding to the oblivion and invisibility to which Vajont had been condemned. Perhaps more than the permanent exhibition, the museum's reference to "participatory memory" proposes an important insight. It cites the annual meeting of survivors as an example of this, taking place on the Sunday preceding the anniversary of the disaster. If it is true that the memories of Vajont can only be shared and divided at the same time, then what is needed is not just a traditional museum but a laboratory capable of animating the citizen humanities[21] we often discuss—broad cultural mobilizations that evoke the ghosts of the past within the places and stories of our present.

CONCLUSION

I have been keeping an appointment with Vajont since 1997. It was by no means certain that I would eventually meet the dam in reality. Twenty-six years ago, my future was uncertain. Always chasing a salary, I wandered between unlikely job openings and the dream of leaving Italy for good. Eventually, I did leave, and it was from afar that I reencountered the Vajont Dam. Of course, geographical distance delayed that encounter, since it is not easy to tell a story like this without the opportunity to experience it from within. Yet, perhaps it was also the distance that helped me return to Vajont, to see things in it that I would not have seen had I not gone far away and to tell it as I have tried to do in this book.

The journey to Vajont was long and did not exactly follow the fastest route. It passed through the United States—Kansas, Connecticut, California—and through the discovery of American environmental history, which provided a release from certain cul-de-sacs in which the Italian debate seemed stranded.

The Columbia River does not pass through Belluno, yet I could not have read the Vajont valley as a hydroelectric laboratory without passing through Richard White's *Organic Machine*.[1] Venice is not Los Angeles, but the subaltern positionings that place a valley in the service of the metropolis are best seen—unpredictably—from the hydraulic empire of the American West, recounted by Donald Worster.[2] The modernist project and its carryover of ecological and scientific simplification—a valley becoming an engine—are best glimpsed from New Haven through chatting with James Scott.[3] This long detour to Vajont led me to Dixie, Texas, and from there to the Land of Fires in Campania. The process pushed me beyond the disciplinary boundaries within which I had previously operated as a disciplined historian, bringing me into dialogue with political ecologists and other scholars who question the unjust distribution of environmental risks and the politicization of both scientific and popular knowledge. Joan Martinez, Laura Pulido, David Pellow, Robert Bullard, and Julie Sze have all been excellent guides on this convoluted and beautiful journey to Vajont.[4] Viewed from Stockholm, however, the valley was full of stories. The environmental humanities made me discover not only the power of the narratives inscribed within the rocks of the Toc, the concrete of the dam, and the gravestones of the cemetery but also the geological relationships and the voices of the survivors. Without Rob Nixon, Serenella Iovino, and Stefania Barca, I would not have made it to my meeting with Vajont—at least not with the Vajont that I was able to meet in some part thanks to them.[5]

When I first pledged myself to the Vajont story in 1997—unaware of where it would lead me—the Anthropocene had yet to arrive. Or, rather, it had perhaps already been with us for a few decades or even centuries, but no one had named it yet. One cannot return to Vajont without keeping this global narrative in mind, and its ambition to explain both our history and our future. As I mentioned in the introduction, Vajont can be seen as a perfect parable for the Anthropocene: a story of human hubris—of believing that technology can bend the laws of nature—and of our own smallness in the face of nature's power, a reality brutally underscored by the disaster's tragic conclusion.

If I presented this notion in the introduction, by the end of the journey through Vajont, I hope it has become clear that this is not my reading of the story. Rather than speaking simply of the Anthropocene, Vajont tells us a story of the Capitalocene, or perhaps the Wasteocene, as I have termed it. At Vajont, it is not humans generically abusing nature but a particular way of conceiving and organizing socioecological relations that determines who and what is expendable in the name of profit. As always in the Wasteocene, the imposition of wasting relations—at Vajont taken to the extreme with the slaughter of two thousand people—is accompanied by the construction of a toxic narrative infrastructure that invisibilizes and naturalizes the disaster while establishing hierarchies and criteria of legitimacy between knowledge and regimes of memory. As Stefania Barca has written, we must dismantle the mainstream narrative of the Anthropocene to achieve narrative justice—one that allows us to recognize those killed by the project of modernization and progress, whether in the Vajont valley or in the Amazon.[6]

The world seen from Vajont—our past and our future—can certainly be frightening, but it can also offer hope. Nothing about Vajont was scripted or inevitable. It did not require science fiction–level technological innovations, nor was science particularly incapable of understanding the phenomena involved. Clearly, at Vajont, there were uncertainties, but these could have been used as reasons for caution, not as excuses for taking gambles. In the aftermath of the disaster, Tina Merlin declared, "I wish I had disturbed public order!" Let's act so that the same regrets will not torment us.

NOTES

PROLOGUE

1. The Adriatic Electricity Company was founded in the early 1900s by Giuseppe Volpi and Count Ruggero Revedin. Volpi was also a prominent political figure during the fascist regime, serving first as the governor of Libya from 1921 to 1925 and then as the minister of finance from 1925 to 1928.

INTRODUCTION

1. Michel-Rolph Trouillot, *Silencing the Past: Power and the Production of History* (Boston: Beacon Press, 1995); and bell hooks, *Belonging: A Culture of Place* (New York: Routledge, 2008).

2. Geordie is a traditional British folk ballad, telling the story of a poacher who is condemned to die because he had killed six deer in a royal reserve.

3. Marco Armiero, *A Rugged Nation: Mountains and the Making of Modern Italy* (Cambridge: White Horse Press, 2011).

4. Italian historians speak of an "economic miracle" to describe the years between the 1950s and 1960s, when the country experienced rapid industrialization and economic growth.

5. Silvio O. Funtowicz and Jerome R. Ravetz, "Science for the Post-Normal Age," *Futures* 25, no. 7 (1993): 739–755.

6. Sandra Harding, "Strong Objectivity: A Response to the New Objectivity Question," *Synthese* 104, no. 3 (1995): 331–349.

7. In 2016, the International Commission on Stratigraphy declined to formally recognize the Anthropocene as a new geological epoch due to several reasons, including the lack of a clear stratigraphic signal, known as the so-called golden spike.

8. Marco Armiero, *Wasteocene: Stories from the Global Dump* (Cambridge: Cambridge University Press, 2021).

9. I am referring here to the Gospel quote: "It is easier for a camel to go through the eye of a needle than for a rich man to enter the kingdom of God" (Matthew 19:24).

CHAPTER 1

1. Micaela Coletti, Testimony, accessed August 25, 2025, http://www.sopravvissutivajont.org/storia_micaela.asp.

2. *Vajont. La tragedia. Sopravvissuti, superstiti, testimoni, immagini*, adapted from an idea of Titano Pisani, on the advice of Stefano Gambarotto, CafèTv24, October 11, 2009, https://www.youtube.com/watch?v=6Kbhq7Wq5gY.

3. Renzo Scagnet, Testimony, accessed August 25, 2025, http://www.sopravvissutivajont.org/storia_renzo.asp.

4. Matelda Coletti, Testimony, accessed August 25, 2025, http://www.sopravvissutivajont.org/storia_matelda.asp.

5. *Vajont. La tragedia. Sopravvissuti, superstiti, testimoni, immagini*, https://www.youtube.com/watch?v=6Kbhq7Wq5gY&t=2707s.

6. Federico Ferraù, *9 ottobre 1963. Il superstite: vi racconto il mio Vajont*, *Il Sussidiario*, October 9, 2013, https://www.ilsussidiario.net/news/cronaca/2013/10/9/9-ottobre-1963-il-superstite-vi-racconto-il-mio-vajont/433474/.

7. Officine Tolau, *Cinque ore prima del Vajont*, May 11, 2011, https://www.youtube.com/watch?v=-J5e43ZObzo.

8. Officine Tolau, *Cinque ore prima del Vajont*.

9. Giuseppe Vazza, *In meno di quattro minuti: testimonianza sul Vajont* (Padua: CLEUP, 2017), 46.

10. Francesco Sossai, *"Ricordo come fosse ieri," il racconto dei superstiti del Vajont*, *Il Corriere delle Alpi*, August 31, 2013, https://video.corrierealpi.gelocal.it/dossier/vajont-1963-2013/ricordo-come-fosse-ieri-il-racconto-dei-superstiti-del-vajont/17045/17082.

11. Russell Eatough, dir., *Seconds from Disaster. Mountain Tsunami*, National Geographic, 2012, https://www.youtube.com/watch?v=IY78pbrJrmw.

12. Fondazione Vajont, "Ricordando quella notte. Testimonianza di Vittore Zannol," *Vajont 1963–2013*, May 22, 2013, https://www.youtube.com/watch?v=gzw4gN2OtF0.

13. Vazza, *In meno di quattro minuti*, 32.

14. Elsa Capraro, "I miei ricordi del Vajont, in Marina Zerman," *La voce delle donne*, June 3, 2013, https://www.youtube.com/watch?v=-DhoFR0YLY4&t=702s.

15. Vazza, *In meno di quattro minuti*, 33–34.

16. Andrea Pais, *Longarone*, *Antincendio e Protezione Civile Rivista Mensile*, n. 59, 1963, https://www.anvvftorino.com/uploads/6/1/4/4/61440757/l_opera_del_corpo_vv.f._nel_vajont.pdf

17. Stefano Campolo collected Zanarini's testimony in his article "Gli operai di servizio si salvarono, quelli di riposo no," *Corriere delle Alpi*, speciale Vajont 1963–2013, the link is no longer accessible; I last accessed it in June 2023. , http://temi.repubblica.it/corrierealpi-diga-del-vajont-1963-2013-il-cinquantenario/non-capivo-se-stavo-sognando-o-se-era-lapocalisse/.

18. Pais, *Longarone*.

19. Campolo, "Gli operai di servizio si salvarono."

20. Campolo, "Gli operai di servizio si salvarono."

21. The line "And then we emerged to see the stars once more" is from Dante's Inferno and marks the exit of the poet and his guide, Virgil, from Hell.

22. Italico Cauteruccio, *Intervento degli Alpini nella tragedia del Vajont*, accessed 25 August 2025, https://sites.google.com/site/mulidellacadore/racconti-e-poesie/intervento-degli-alpini-nella-tragedia-del-vajont.

23. "I cadaveri erano anche sugli alberi", interview with Giovanni Boite Stella, *Corriere delle Alpi* speciale Vajont 1963–2013, August 31, 2013, https://video.corrierealpi.gelocal.it/dossier/vajont-1963-2013/vajont-1963-i-cadaveri-erano-anche-sugli-alberi/17040/17076. The link is no longer accessible; I last accessed it in June 2023.

24. Viviana Capraro, ed., *L'abbraccio e la parola. Testimonianze dei soccorritori del Vajont* (Longarone: Grafiche Longaronesi, 2013), 44.

25. Capraro, *L'abbraccio e la parola*, 66.

26. Felice Borsato, *Longarone Anno Zero* (Rome: Editrice la Cronistoria, 1963), 10–11.

27. Capraro, *L'abbraccio e la parola*, 17.

28. Interview by the author with Mr. Angelo Caponnetto, November 16, 2022.

29. Borsato, *Longarone anno zero*, 5.

30. Alessandro Feliziani, "Vajont, 50 anni dopo. La testimonianza di un alpino," *Cronache Maceratesi*, October 9, 2013, https://www.cronachemaceratesi.it/2013/10/09/vajont-50-anni-dopo-la-testimonianza-di-un-alpino/384861/.

31. Ulderico Quintabà, "Il Vajont dei soccorsi," *Oggi Treviso*, October 9, 2014, https://www.oggitreviso.it/vajont-dei-soccorsi-97902.

32. Interview by the author with Enrico Biagioli. September 28, 2022.

33. Interview by the author with Enrico Biagioli.

34. Interview by the author with Enrico Biagioli.

35. Pais, *Longarone.*

36. Cauteruccio, *Intervento degli Alpini.*

37. Vazza, *In meno di quattro minuti*, 63.

38. Lucia Vastano, *Vajont, l'onda lunga* (Florence: Ponte delle Grazie, 2008).

39. Lucia Vastano, "La memoria offesa," *Narcomafie*, June 1, 2002, http://www.archivio900.it/it/articoli/art.aspx?id=7907.

40. Paolo Rumiz, "Lo spettro della corriera fantasma tormenta la memoria del Vajont," *La Repubblica*, June 17, 2002, https://ricerca

.repubblica.it/repubblica/archivio/repubblica/2002/06/17/lo-spettro-della-corrierafantasma-tormenta-la-memoria.html.

41. An unsigned article published on the Tempi.it portal claimed that the woman in question was a sex worker who worked along the state highway. No one recognized her, not only because she was not from Longarone but also to avoid having to admit to having known her. Tempi, Il bosco del Vajont, October 13, 2003, accessed August 25, 2025, https://www.tempi.it/il-bosco-del-vajont/.

42. In Paolo Rumiz's version of the story, Alvise did not tell of the discovery because he was carrying out unauthorized digging in search of metals for resale.

CHAPTER 2

1. Richard White, *The Organic Machine* (New York: Hill and Wang, 1995), ix.

2. Claudio Datei, *Vajont: Una storia idraulica* (Padua: Cortina, 2005), 8.

3. Pietro Ciarlo, "Il testo unico del 1933 sulle acque e sugli impianti elettrici," in *Storia dell'industria elettrica in Italia, vol. iii, Espansione e oligopolio. 1926–1945*, ed. Giuseppe Galasso (Rome: Laterza, 1993), 61.

4. Giacomo Bonan, *Le acque agitate della patria l'industrialializzazione del Piave (1882–1966)* (Rome: Viella, 2020), 74.

5. James Sievert, *The Origins of Nature Conservation in Italy* (Bern: Peter Lang, 2000), 90.

6. Franco Saverio Nitti, *La Conquista della Forza* (Turin: Roux-Viarengo, 1905).

7. James Scott, *Seeing Like a State: How Certain Schemes to Improve the Human Condition Have Failed* (New Haven, CT: Yale University Press, 1998), 88–89.

8. Bonan, *Le acque agitate della patria*, 134.

9. Letter from Carlo Semenza to Vincenzo Ferniani, April 20, 1961, in Edoardo Semenza, *La storia del Vajont raccontata al geologo che ha scoperto la frana* (Ferrara: K-flash, 2001), 227.

10. Maurizio Reberschak, ed., "Una storia del 'genio italiano': il Grande Vajont," in *Il Grande Vajont* (Sommacampagna: Cierre, 2003), 23–57.

11. Eleonora Pierobon, "Disastro del Vajont: un modello sph per l'analisi dei risultati sperimentali sui modelli fisico-idraulici" (master's thesis, University of Padua, 2013–2014), 8.

12. Marco Paolini and Gabriele Vacis, *Il racconto del Vajont* (Milan: Garzanti, 1997; repr. 2021). I will return more fully to Paolini's theatrical work in chapter 5.

13. The Consiglio Superiore dei Lavori Pubblici (Italian Higher Council of Public Works) is a technical advisory body within Italy's Ministry of Infrastructure and Transport. It provides expert evaluations on major public works projects and oversees the quality and safety of these key infrastructures.

14. As Maurizio Reberschak, undoubtedly the historian most dedicated to Vajont, recalled, this further expansion of the project was made possible by a government measure in November 1956 that supported the expansion of hydroelectric infrastructure. In Reberschak, "Una storia del 'genio Italiano,'" 30.

15. Semenza, *La storia del Vajont*, 32.

16. Pierobon, *Disastro del Vajont*, 9.

17. Giorgio Dal Piaz, "Relazione geologica su una sezione della valle del Vajont presa in esame per una diga di sbarramento idraulico" (Geological report on a section of the Vajont Valley considered for a hydraulic dam), August 7, 1937, in *Commissione Parlamentare di Inchiesta* (CPI; Parliamentary Commission of Inquiry), Working Group 1, 2.

18. Enciclopedia Treccani, "DAL PIAZ, Giorgio," accessed August 25, 2025, https://www.treccani.it/enciclopedia/giorgio-dal-piaz_%28Enciclopedia-Italiana%29/.

19. Letter from Giorgio Dal Piaz to Carlo Semenza, February 6, 1957. This document has been published many times. I quote it here from Semenza, *La storia del Vajont*, 147.

20. On Dal Piaz's letter, Edoardo Semenza writes: "I think that between honest people and friends it can be conceived that relations

such as these are not suspect: there is no shadow of servility, but only a request to receive a text that he had liked"; Semenza, *La storia del Vajont*, 147.

21. CPI, Working Group 1, 2–3.

22. Semenza, *La storia del Vajont*, 147.

23. CPI, Working Group 1, 4.

24. CPI, Working Group 1, 10.

25. CPI, Working Group 1, 14.

26. Leopold Müller, "Rapporto geologico preparato per conto della SADE" (Geological report prepared on behalf of SADE), February 3, 1961, in CPI, Working Group 1, XXXIX/1729, 13.

27. Funtowicz and Ravetz, "Science for the Post-Normal Age."

28. Francesco Penta, "Promemoria sulle indagini nella zona del Toc e nelle vicinanze dell'abitato di Erto" (Memo on the investigations in the Toc area and in the vicinity of the town of Erto), Rome, October 31, 1961, in CPI, Working Group 1, 4.

29. This quotation from a memo sent by Penta to the chairman of the testing commission can be found in CPI, Working Group 1, 16.

30. Pierobon, *Disastro del Vajont*, 27–28.

31. Fiorello Zangrando, "Ghetti: 'Con il mio modello, si poteva prevedere il disastro,'" *Il Gazzettino*, March 11, 1969, http://www.sopravvissutivajont.org/jshow.asp?id=75.

32. Andrea Rinaldo, "Commemorazione del socio effettivo Claudio Datei," in *Atti dell'Istituto Veneto di Scienze, Lettere ed Arti*, Tomo XX (2012–2013), Classe di scienze fisiche matematiche e naturali, n.p. ma p. 4, accessed August 25, 2025, http://www.istitutoveneto.org/pdf/commemorazione_datei.pdf.

33. Zangrando, "Ghetti."

34. This passage from the Ghetti reports is cited by Datei, *Vajont*, 32.

35. Datei, *Vajont*, 48.

36. Datei, *Vajont*, 95.

37. Datei, *Vajont*, 35.

38. Quoted, unsourced, from Pierluigi Favero, *Ipotesi sul Vajont* (Belluno: Aics, 2021), 126.

39. Vastano, *Vajont, l'onda lunga*, 158–159.

40. Vazza, *In meno di quattro minuti*, 26–27.

41. Alessandra Mercanzin, "Sopravvivere al Vajont. Intervista a Giuseppe Vazza," *7inpunto*, November 5, 2014, https://www.youtube.com/watch?v=Z64DzxGvHTQ https://www.youtube.com/watch?v=-Al9XI5G7ps.

42. Vazza, *In meno di quattro minuti*, 28–29.

43. "At home, I had then told my wife, my mother, and the customers who were in the store everything"; Vazza, *In meno di quattro minuti*, 26.

44. Vazza, *In meno di quattro minuti*, 28.

45. "Storie e Persone. Memorie di Comunità—La tragedia della diga del Vajont," April 15, 2008, https://www.youtube.com/watch?v=Z64DzxGvHTQ.

46. Gianni Minoli, "Mixer," May 6, 1996, https://www.youtube.com/watch?v=zOTbSMe1Ej4.

47. Parliamentary Commission of Inquiry into the Vajont Disaster, *Relazione al Ministro dei Lavori Pubblici* (Rome: Istituto Poligrafico dello Stato, 1964), 58.

48. "La Storia Siamo Noi Dossier Vajont Una Tragedia Annunciata," accessed August 25, 2025, https://www.youtube.com/watch?v=zOTbSMe1Ej4 (Corona's testimony starts at minute 40).

49. Renato Zanolli, *Vajont, Cronaca di una tragedia annunciata* (Vittorio Veneto: De Bastiani, 2013), 52.

50. Lorenzo Barutta, *Vajont diario di un sopravvissuto*, accessed August 25, 2025, https://www.youtube.com/watch?v=C_0YkjUXLXY. This is a short film awarded the *La memoria in corto* prize by the Alfredo Galmozzi Research Center.

51. Officine Tolau, *Cinque ore prima del Vajont*.

52. Vazza, *In meno di quattro minuti*, 31.

53. Eatough, *Seconds from Disaster*.

54. Giorgio Tosi, *Vajont: mors inimica venit* (Padua: CLEUP, 2009), 12–14.

55. Eduardo Frittoli, "Diga di Ponesei: 60 anni fa l'incidente che anticipò il Vajont," *Panorama*, March 22, 2009, https://www.panorama.it/news/diga-pontesei-1959-incidente-vajont; and Enrico Bacchetti, "100 anni, 4 minuti, 1910 morti. Breve viaggio nella tragedia del Vajont," accessed August 25, 2025, http://www.novecento.org/dossier/italia-didattica/100-anni-4-minuti-1910-morti-breve-viaggio-nella-tragedia-del-vajont/.

56. The letter is quoted in CPI, Working Group 1, 16.

57. Minoli, "Mixer," https://www.youtube.com/watch?v=zOTbSMe1Ej4 41:51.

58. From a report by Italian Public Television in the immediate aftermath of the disaster; https://www.youtube.com/watch?v=Ryw4Qy0ZSYo 3:25.

CHAPTER 3

1. In ancient Greek mythology, Cassandra was a prophetess cursed to utter true prophecies that no one would believe. Toni Sirena, Tina Merlin's son, mentioned his mother's discomfort with this mythological and fatalistic reference, which framed the disaster as a predetermined destiny, turning her work as a journalist into a sort of divination.

2. "Procura della Repubblica di Milano, "Citazione in giudizio per direttissima, 30 novembre 1960, Procedimento a carico di Pizzigoni Orazio e Merlin Clementina" (Milan Public Prosecutor's Office, Summons for summary judgment, November 30, 1960, Proceedings against Pizzigoni Orazio and Merlin Clementina), in State Archives of Belluno. My gratitude to Adriana Lotto, who provided me with a reproduction of this bundle.

3. Tina Merlin, "La Sade spadroneggia ma i montanari si difendono," *L'Unità*, May 5, 1959, http://temi.repubblica.it/corrierealpi-diga

-del-vajont-1963-2013-il-cinquantenario/la-sade-spadroneggia-ma-i-montanari-si-difendono/.

4. Zanolli, *Vajont, Cronaca di una tragedia annunciata*, 25.

5. Minoli, "Mixer," https://www.youtube.com/watch?v=zOTbSMe1Ej4.

6. Tribunale di Milano, sezione 3a penale, udienza 10/11/1960 (Tribunal of Milan, sec. 3a penal, hearing 10/11/1960), in State Archives of Belluno.

7. Tribunale Civile e Penale di Milano, sezione penale, sentenza numero 1974, 15 dicembre 1960 (Civil and Criminal Court of Milan, criminal section, judgment number 1974, December 15, 1960), in State Archives of Belluno.

8. Tribunale Civile e Penale di Milano, sezione penale, sentenza numero 1974, 15 dicembre 1960.

9. Tina Merlin, "Magari fossi riuscita a turbare l'ordine pubblico," *L'Unità*, October 13, 1963, https://www.tgvallesusa.it/vajont-magari-fossi-riuscita-a-turbare-lordine-pubblico/.

10. Adriana Lotto, *Quella del Vajont: Tina Merlin, una donna contro* (Sommacampagna: Cierre, 2011).

11. Armiero, *Wasteocene*.

12. Semenza, *La storia del Vajont*, 141.

13. The philosopher Donna Haraway described this claim to objectivity as a god trick: Like god, the researcher positions themselves outside the context they are studying, assuming that knowledge can only be obtained by adopting an impartial, objective, and entirely disembodied point of view. Donna Haraway, "Situated Knowledges: The Science Question," *Feminist Studies* 14, no. 3 (1988): 575–599.

14. Tina Merlin, *Sulla pelle viva. Come si costruisce una catastrofe. Il caso Vajont* (Sommacampagna: Cierre, 2001), 21.

15. Tina Merlin, "Una gigantesca frana precipita a Erto nel lago artificiale costruito dalla SADE," *L'Unità*, November 8, 1960, https://tinamerlin.it/pubblicazioni/articoli-di-tina-merlin/vajont/.

16. Merlin, "Una gigantesca frana."

17. Tina Merlin, "Il crollo della diga del Vajont," *L'Unità*, October 10, 1963, https://tinamerlin.it/pubblicazioni/articoli-di-tina-merlin/vajont/.

18. Interview by the author with Mr. Toni Sirena, November 21, 2022.

19. Tina Merlin, "*L'Unità* fu processata per aver denunciato il pericolo," *L'Unità*, October 11, 1963, https://tinamerlin.it/pubblicazioni/articoli-di-tina-merlin/vajont/.

20. Interview by the author with Mr. Toni Sirena cit. Benigno Zaccagnini (1912–1989) was a relevant politician in the Christian Democratic Party.

21. Merlin, *Sulla pelle viva*, 180.

22. Sandro Canestrini, *Vajont: genocidio dei poveri* (Verona: Cierre edizioni, 2003), 29–30.

23. The expression "in Agramante's camp" can be translated as "in the enemy's camp." It implies the disadvantage of fighting under conditions and rules that favor one of the contenders. The expression originates from the poem *Orlando Furioso*, written by Ludovico Ariosto in the sixteenth century. In Ariosto's fantastic world, Agramante was a Muslim king fighting against Charlemagne's knights.

24. The framework of *legittima suspicione* is a legal measure through which a trial is removed from the jurisdiction of the geographically competent court, due to the concern that the judges may be excessively influenced by the local context, thereby compromising the impartiality and fairness of their judgment.

25. Dino Tonini was the director of the SADE study office; Roberto Marin was the general manager of SADE; Augusto Ghetti was the director of the Institute of Hydraulics and Hydraulic Constructions at the University of Padua and was in charge of Nove's experiments on landslide effects in the Vajont basin; Pietro Frosini was a member of the testing commission; and Almo Violin was the director of the Belluno Civil Engineering Department.

26. ENEL (originally Ente Nazionale per l'Energia Elettrica) was founded in 1962 as a state-owned entity. It acquired SADE and all other private energy producers.

27. Wikipedia, "Alberico Biadene," accessed August 25, 2025, https://it.wikipedia.org/wiki/Alberico_Biadene.

28. La bilancia della giustizia e Rubò 55 lire. Due anni di carcere, Lotta Continua, May 22, 1972, 2, http://www.lalottacontinua.it/giornale-archivio/LC1_1972_05_21.pdf.

29. Canestrini, *Vajont: genocidio dei poveri*, 30.

30. Fiorello Zangrando, "Il lungo viaggio attraverso la colpa," in *Il Grande Vajont*, ed. Maurizio Raberschak (Sommacampagna: Cierre, 2003), 233.

31. Judge Mario Fabbri recalled the technical and administrative challenges faced by the Belluno prosecutor's office in acquiring and cataloging 5,200 seized documents, many of which included numerous attachments. See Mario Fabbri, "Processo penale e inchiesta parlamentare. Rapporti, interferenze, conflitti tra autorità giudiziaria e commissione," in *Il Grande Vajont*, ed. Maurizio Raberschak (Sommacampagna: Cierre, 2003), 183.

32. Canestrini, *Vajont: genocidio dei poveri*, 51.

33. Dino Buzzati (1906–1972) was an Italian journalist and writer. His most well-known novel is *Il deserto dei Tartari* (1940), translated as *The Tartar Steppe* and *The Stronghold*. As a journalist, he worked his entire life with the renowned newspaper *Il Corriere della Sera*. Giorgio Bocca (1920–2011) was also a writer and journalist, working for the Italian newspaper *La Repubblica*. Bocca is best known for his work on the Italian Resistance against the Nazi-fascist regime, as he was a partisan himself.

34. Historian Mario Isnenghi analyzed how various journalistic outlets of the time were divided between attributing blame and offering an all-naturalistic explanation for the disaster. Mario Isnenghi, "Il potere di carta. Il dopo-Vajont e la lotta delle parole," in *Il Grande Vajont*, ed. Maurizio Reberschak (Sommacampagna: Cierre, 2003), 101–165.

35. The committee included Professors Michele Gortani, Ardito Desio, and Joos Cadich (geologists), Carlo Morelli (geophysicist), and Francesco Ramponi and Giulio Citrini (engineers).

36. Pierobon, *Disastro del Vajont*, 35.

37. This second committee consisted of Floriano Calvino, Marcel Roubault, Henri Gridel, and Alfred Stucky. Due to his engagement in the trial, Calvino lost his position as a temp at the University of Padua.

38. Flavio Zanonato, "In ricordo del compagno Giorgio Tosi," March 13, 2021, https://www.facebook.com/notes/10225002582680499/.

39. Tosi, *Vajont: mors inimica venit*, 10–11.

40. Tosi, *Vajont: mors inimica venit*, 11–12.

41. Odoardo Ascari, *Una arringa per Longarone* (Feltre: P. Castaldi, 1973), PDF p. 6 https://it.scribd.com/document/78307398/Un-Arringa-Per-Longarone-odoardo-Ascari

42. Procura della Repubblica di Belluno, Requisitoria nel procedimento penale contro Biadene, Pancini et a., P.M. Arcangelo Mandarino, Belluno 22 novembre 1967, N.818/63 P. M., N. 85/64 G. I. (Prosecutor's Office of Belluno, Requisition in the criminal proceedings against Biadene, Pancini et a., P.M. Arcangelo Mandarino, Belluno November 22, 1967, No. 818/63 P. M., No. 85/64 G. I.), in Canestrini Archives 1.41, 137, preserved at the Biblioteca Civica B. Tartarotti Rovereto.

43. Requisitoria del Pubblico Ministero, 22 novembre 196—Busta 34 (Prosecutor's memorandum, November 22, 1967—Envelope 34).

44. Canestrini, *Vajont*, 99.

45. Canestrini, *Vajont*, 97–99.

46. Canestrini Archives 1.41, 46.

47. Procura della Repubblica di Belluno, Requisitoria del PM Arcangelo Mandarino, Belluno 22 Novembre 1967, N. 818/63 P.M. (Prosecutor's Office, Closing Argument of Public Persecutor

Arcangelo Mandarino, Belluno November 22, 1967, No. 818/63 P.M.), in Canestrini Archives 1.41, 47.

48. Sentenza n. 112, 17 dicembre 1969 (Judgment No. 112, December 17, 1969), in Canestrini Archives 1.44b, 261.

49. Procura della Repubblica di Belluno, Requisitoria del PM Arcangelo Mandarino, Belluno 22 Novembre 1967, N. 818/63 P.M. (Belluno Prosecutor's Office, Closing argument of Public prosecutor Arcangelo Mandarino, Belluno November 22, 1967, No. 818/63 P.M.), in Canestrini Archives 1.41, 131.

50. Avv. Giovanni Carloni, Arringa per le parti civili, Corte di Appello degli Abruzzi (Closing argument for civil parties by the lawyer Giovani Carloni, Abruzzi Court of Appeal), in Canestrini Archives 1.41, 35.

51. Procura della Repubblica di Belluno, Requisitoria del PM Arcangelo Mandarino (Belluno Prosecutor's Office, Closing argument of Public Prosecutor Arcangelo Mandarino, Belluno November 22, 1967, N. 818/63 P.M.), in Canestrini Archives, 1.41, 88.

52. Carloni, Closing argument for civil parties, in Canestrini Archives 1.41, 59.

53. Judgment No. 112, December 17, 1969, in Canestrini Archives 1.44a, 130.

54. Archivio di Stato di Belluno, Archivio del processo penale del Vajont, Tribunale di Belluno. Ufficio Istruzione, b. 34, Requisitoria del Public Prosecutor Mandarino, 22 novembre 1967 (State Archives of Belluno, Archives of the Vajont criminal trial, Court of Belluno. Office of Preliminary Investigations, b. 34, Closing argument of Public Prosecutor Mandarino, November 22, 1967).

55. Archivio di Stato di Belluno, Archivio del processo penale del Vajont, Tribunale di Belluno. Ufficio Istruzione, b. 29, fasc. 2b, pp. 1–104: Prima memoria dell'avvocato Bras in difesa dell'ingegnere Biadene del 10.05.1966 (State Archives of Belluno, Archives of the Vajont criminal trial, Court of Belluno. Office of Preliminary Investigations, b. 29, fasc. 2b, pp. 1–104: First brief of lawyer Bras in defense of engineer Biadene dated May 10, 1966).

56. As I have outlined above, Penta envisioned two possible scenarios regarding the landslide: one more optimistic, the other catastrophic.

57. Archivio di Stato di Belluno, Archivio del processo penale del Vajont, Tribunale di Belluno. Ufficio Istruzione, b. 29, fasc. 2e, pp. 1–61 e allegati: Nota difensiva degli avvocati De Marsico-Ottolenghi nell'interesse dell'ingegnere Pancini del 25.08.1966 (State Archives of Belluno, Archives of the Vajont criminal trial, Court of Belluno. Office of Preliminary Investigations, b. 29, fasc. 2e, pp. 1–61 and attachments: Defense memorandum by lawyers De Marsico-Ottolenghi in the interest of engineer Pancini dated 25 August 1966).

58. Carloni, Closing argument for civil parties, 28.

59. Tosi, *Vajont: mors inimica venit*, 24.

60. Procura Generale della Repubblica presso la corte di appello degli Abruzzi, Motivi di ricorso per cassazione proposto dal PM avverso la sentenza 3/10/1970 (Prosecutor General's Office at the Abruzzi Court of Appeals, Grounds of Appeal brought by the Public Prosecutor against the judgment 3 October 1970), PDF p. 20 (unpaginated), in Canestrini Archive 1.41, grounds of appeal.

61. Procura Generale della Repubblica presso la corte di appello degli Abruzzi, Motivi di ricorso per cassazione proposto dal PM avverso la sentenza 3/10/1970 (Prosecutor General's Office at the Abruzzi Court of Appeals, Grounds of Appeal brought by the Public Prosecutor against the judgment 3 October 1970, PDF p. 20 (unpaginated), in Canestrini Archive 1.41, grounds of appeal.

62. Archivio di Stato di Belluno, Archivio del processo penale del Vajont, Tribunale di Belluno. Ufficio Istruzione, b. 2, all. B, fascicolo 2, foliazione p. 379 e ss.: Interrogatorio dell'ingegnere Biadene del 14 novembre 1963 (State Archives of Belluno, Vajont criminal trial archives, Court of Belluno. Office of Preliminary Investigations, b. 2, all. B, file 2, foliation p. 379 ff: Interrogation of engineer Biadene of November 14, 1963).

63. Archivio di Stato di Belluno, Archivio del processo penale del Vajont, Tribunale di Belluno. Ufficio Istruzione, b. 2, all. B, fascicolo

2, foliazione p. 379 e ss.: Interrogatorio dell'ingegnere Biadene del 14 novembre 1963 (State Archives of Belluno, Vajont criminal trial archives, Court of Belluno. Office of Preliminary Investigations, b. 2, all. B, file 2, foliation p. 379 ff: Interrogation of engineer Biadene of November 14, 1963).

64. Mario Passi, "Assolto Lorenzo Rizzato per il documento SADE," *L'Unità*, October 22, 1963, https://archivio.unita.news/assets/main/1963/10/22/page_003.pdf.

65. Cristiano Cadoni, "'Così rubai le prove all'Università,' Rizzato racconta dopo 38 anni dal suo gesto nei giorni del Vajont," *Corriere delle Alpi*, November 7, 2001, http://www.sopravvissutivajont.org/jshow.asp?id=66.

66. Procura della Repubblica presso il tribunale di Padova, Procedimento penale contro Rizzato Lorenzo (Public Prosecutor's Office at the Court of Padua, Criminal Proceedings against Rizzato Lorenzo), 184/64.

67. Archivio Tribunale Padova, Verbale confronto tra Rizzato e Benedini (Padua court archives, Minutes of confrontation between Rizzato and Benedini).

68. Sentenza Rizzato (Rizzato judgment), no page but 17.

69. Interview by the author with Mr. Pierantonio Rizzato, May 8, 2023.

70. Mario Passi, "Arrestato per un documento che compromette la SADE," *L'Unità*, October 16, 1963.

71. Actually, another source identifies the MP Giuseppe Golinelli, also from the PCI, as the one who accompanied Mario Passi to the prosecutor's office in Belluno (*Rivista di giurisprudenza penale*, *Il Foro Italiano* 94, no. 3 [1971]: 204). This is a minor discrepancy since the operation was undoubtedly led by MP Busetto.

72. Rizzato judgment, PDF p. 20 (unpaginated).

73. Interview by the author with Mr. Pierantonio Rizzato, May 8, 2023.

74. Fabrizio Zupo, "Il teatro e le rivelazioni sul disastro Vajont, addio all'eroe nascosto Lorenzo Rizzato, Il mattino di Padova," July 29,

2019, https://mattinopadova.gelocal.it/padova/cronaca/2019/07/29/news/il-teatro-e-le-rivelazioni-sul-disastro-vajont-addio-all-eroe-nascosto-lorenzo-rizzato-1.37254755.

CHAPTER 4

1. Armiero, *Wasteocene*.

2. Giovanni De Luna, *La repubblica del dolore. Le memorie di un'Italia divisa* (Milan: Feltrinelli, 2011).

3. Daniele Giglioli, *Critica della vittima: un esperimento con l'etica* (Rome: Edizioni Nottetempo, 2015).

4. Andrea Suverato, *Scritture complici: uno studio sulle finzioni testimoniali di inizio millennio*, (PhD diss., University of Bologna), pp. 83–84.

5. Taranto hosts the largest steel factory in Europe, accused of polluting the entire city, especially the Tamburi district. Terra dei Fuochi is an area between the cities of Naples and Caserta, characterized by toxic fires and illegal dumping. Gela is a petrochemical town in Sicily with high incidences of cancers, stillbirths, malformations, and other diseases connected with industrial contamination.

6. I developed the concept of toxic narrative infrastructure as a set of narratives that render invisible, normalize, or naturalize injustice in my *Wasteocene*, 18–26.

7. Vastano, *Vajont, l'onda lunga*.

8. Lucia Vastano, interview with Carolina, 2017, part of the Toxic Bios project. Toxic Bios is a public environmental humanities project I directed, aimed at collecting personal stories of environmental injustice.

9. It is perhaps worth noting that as late as 2005, in an interview with the daily newspaper *Libero*, Mauro Corona reiterated the same narrative about "jackal communists," peppered with several factual errors, including the accusation that they supported the economic settlement proposed by ENEL to the victims. Francesco Patti, "Sulla tragedia del Vajont dalla sinistra solo speculazioni politiche. Intervista a Mauro Corona," *Libero*, October 1, 2005. A timely response to

Corona's accusations can be found in Peppino Zangrando, "Il PCI e il Vajont," *I Protagonisti* 89 (2005): 100–113.

10. In my book *Wasteocene,* I argue that toxic narratives create a narrative infrastructure that silences or normalizes injustices while blaming affected communities for the very problems that oppress them. I refer to a toxic narrative infrastructure because, like any infrastructure, it is almost invisible; nevertheless, it shapes our understanding of reality and constrains our imagination.

11. In this chapter, I discuss citizens in a general sense. However, I am fully aware that communities were—and still are—neither coherent nor homogeneous units. Although some individuals rebelled, that does not imply that everyone did. Moreover, even among those who resisted, their motivations may have varied profoundly. This is not to mention the many—likely the majority—who viewed the dam as an opportunity for progress or, more simply, as a source of wage labor.

12. I wrote about commons and mountains in my book *A Rugged Nation. Mountains and the Making of Modern Italy* (Cambridge: White House Press, 2011), 76–86.

13. "Erto e Casso: un pugno di montanari contro il gigante elettrico," in *Corriere delle Alpi, speciale Vajont 1963–2013*, http://temi.repubblica.it/corrierealpi-diga-del-vajont-1963-2013-il-cinquantenario/erto-prima/. On this, see also Reberschak, "Una storia del 'genio italiano,'" 31.

14. Toni Sirena confirmed to me that, based on his research at the Genio Civile archive in Belluno, it seems that the Filippin family, including Caterina, had sold their land, although it is unclear whether they obtained particularly favorable conditions.

15. CPI, "Relazione finale comunicata alla Presidenza delle Camere" (Final Report Communicated to the Presidency of the Chambers), July 15, 1965, 169.

16. Merlin, "Magari fossi riuscita a turbare l'ordine pubblico."

17. CPI, "Relazione finale comunicata alla Presidenza delle Camere," 169.

18. CPI, "Relazione finale comunicata alla Presidenza delle Camere," 169.

19. CPI, "Relazione finale comunicata alla Presidenza delle Camere," 170.

20. Merlin, *Sulla pelle viva*, pp. 72–73.

21. The *Carabinieri* is one of Italy's main law enforcement agencies, functioning as both a military and a civilian police force.

22. Mario Passi, *Morire sul Vajont: storia di una tragedia italiana* (Padua: Marsilio, 1968), 33.

23. According to the parliamentary committee, the mountain people either feared for the fate of Erto and Casso or were concerned about the effects of the bypass tunnel. CPI, "Relazione finale comunicata alla Presidenza delle Camere," 170–171

24. Zanolli, *Vajont, Cronaca di una tragedia annunciata*, 24.

25. Barutta, *Vajont diario di un sopravvissuto.*

26. Among the many studies that it would be possible to cite, I recall Ferruccio Vendramini, ed., *Disastro e ricostruzione nell'area del Vajont* (Longarone: Comune di Logarone, 1994); Maurizio Reberschak and Ivo Mattozzi, eds., *Il Vajont dopo il Vajont: 1963–2000* (Venice: Marsilio, 2009); Gianni Silei, *Un banco di prova: la legislazione sul Vajont dalle carte di Giovanni Pieraccini (1963–1964)* (Manduria: Lacaita, 2016); and Oddone Demichelis, Micaela Coletti, and Guido Toffolo, *Psicologia dell'emergenza: il caso Vajont* (Savignano: Artistica Piemontese, 2004).

27. Cortina d'Ampezzo, often simply called Cortina, is a renowned mountain resort town located in the Dolomites in northern Italy.

28. *Vajont 2000 condanne*, dir. Luigi Di Gianni, PCI Film Office, 1970. The documentary is available online in the Audiovisual Archive of the Movimento Operaio e Democratico (Workers' and Democratic Movement).

29. *Vajont 2000 condanne.*

30. Capraro, "I miei ricordi," accessed August 25, 2025, https://www.youtube.com/watch?v=-DhoFR0YLY4&t=702s.

31. Caleidoscopio CIAC, "Longarone: Dimostrazioni di superstiti" (Longarone: Demonstrations by survivors), January 9, 1964, https://patrimonio.archivioluce.com/luce-web/detail/IL5000053801/2/longarone-dimostrazioni-superstiti-5.html.

32. *Vajont 2000 condanne.*

33. Zangrando, "Il lungo viaggio attraverso la colpa," 234.

34. "Libro bianco sulla tragedia del Vajont. Prima documentazione presentata dalla delegazione parlamentare del PCI al Presidente della Repubblica Antonio Segni" (White Paper on the Vajont Tragedy. First documentation submitted by the PCI parliamentary delegation to the President of the Republic, Antonio Segni), October 13, 1963 (Rome: Stabilimento SETI, n.d.), 3.

35. "Libro bianco sulla tragedia del Vajont," 12.

36. "Libro bianco sulla tragedia del Vajont," 13.

37. "Libro bianco sulla tragedia del Vajont," 16–18.

38. Resolution of the PCI Central Committee, Rome, October 14, 1963, "Libro bianco sulla tragedia del Vajont", 67–69.

39. In Italian public discourse, the red regions are those where the PCI held a majority of voters for several decades after 1945: Emilia-Romagna, Tuscany, Umbria, and Marche. In more recent years, however, Umbria and Marche have also witnessed victories by right-wing coalitions.

40. A. Cavallari, "I vivi ricominciano a vivere," *Corriere della Sera*, October 12, 1963.

41. Mauro Scoccimarro, *L'accusa del Vajont. Discorso pronunciato alla Senato della Repubblica nella seduta del 19 ottobre 1963* (The Vajont Indictment. Speech delivered in the Senate of the Republic during the session of October 19, 1963) (Rome: Azienda Tipografica Eredi Dott. Bardi, n.d.), 4.

42. Scoccimarro, *L'accusa del Vajont*, 7.

43. Scoccimarro, *L'accusa del Vajont*.

44. Relazione di minoranza presentata dagli onorevoli Busetto, Vianello, Gaiani, Lizzerò, Scoccimarro, Giaquinto, Vidali e Alicata,

Legislatura IV, Disegni di legge e relazioni—Documenti (Minority report presented by the honorable Busetto, Vianello, Gaiani, Lizzerò, Scoccimarro, Giaquinto, Vidali, and Alicata, 4th Legislature, Bills and Reports—Documents), CPI, 76-bis, 3.

45. Relazione di minoranza, 32–33.

CHAPTER 5

1. Serenella Iovino, *Paesaggio civile* (Milan: Il saggiatore, 2022), 160.

2. Maurizio Reberschak wrote this back in 2009, and unfortunately, things do not seem to have changed since. See Maurizio Reberschak, "Il Vajont dopo il Vajont" in *Il Vajont dopo il Vajont 1963–2000*, ed. Maurizio Reberschak and Ivo Mattozzi (Venice: Marsilio, 2009), 21.

3. Guido Crainz, *Storia del miracolo italiano: culture, identità, trasformazioni fra anni cinquanta e sessanta* (Rome: Donzelli), 1996.

4. Guido Crainz, *Il paese mancato: l'Italia dal miracolo economico agli anni ottanta* (Rome: Donzelli), 2003.

5. Paul Ginsborg, *A History of Contemporary Italy* (London: Penguin Books), 1990.

6. John Dickie, John Foot, and Frank M. Snowden, eds., *Disastro! Disasters in Italy Since 1860: Culture, Politics, Society* (New York: Palgrave), 2002. I do remember, however, the circulation of a call for papers to assemble the contributions for this edited volume; therefore, the absence of a chapter on Vajont reflects more the general amnesia of Italian historiography than the direct responsibility of the editors.

7. Armiero, *Wasteocene*; Donna Houston, "Environmental Justice Storytelling: Angels and Isotopes at Yucca Mountain, Nevada," *Antipode* 45 (2013): 417–435; and Stefania Barca, "Telling the Right Story: Environmental Violence and Liberation Narratives," *Environment and History* 20 (2014): 535–546.

8. I am thinking of the work of the association Cittadini per la Memoria del Vajont (Citizens for the Memory of Vajont), which has fought to obtain a set of measures for symbolic reparation, including

official apologies from the president and the awarding of the Gold Medal of Valor for the victims of the disaster. More recently, an online petition was launched to obtain official apologies from ENEL for what happened in 1963.

9. The Memory of the World project, with its registry containing nearly five hundred documentary collections, aims to promote the preservation and access to documentary sources.

10. Marco Paolini, *Vajont*, dir. Gabriele Vacis, aired on RAI 2, October 9, 1997, available at https://www.youtube.com/watch?v=q7CarpRflzs.

11. Paolini and Vacis, *Il racconto del Vajont*, 101. Whenever possible, I will quote from the script of the monologue.

12. Montanelli was an Italian journalist and writer. During the Fascist occupation of Ethiopia, he "married" a twelve-year-old girl through the colonial practice of *madamato*—a form of sexual enslavement common in the Italian colonies. To the best of my knowledge, Montanelli never expressed remorse for what he had done.

13. *La Domenica del Corriere*, November 10, 1963.

14. Paolini and Vacis, *Il racconto del Vajont*, 122.

15. Renzo Martinelli, dir., *Vajont. La diga del disonore*, Martinelli Film Company International, Rai Cinema, Canal+, Comune di Vajont, 2001.

16. Don Camillo and Peppone are iconic characters from a series of Italian stories by author Giovanni Guareschi, set in a small, fictional village in the Po Valley after World War II. The tales humorously and poignantly explore the ideological tensions of postwar Italy, particularly between the Catholic Church, personified by Father Camillo, and the Communist Party, represented by Peppone, the village's Communist mayor. In the 1970s, the "historic compromise" was an attempt to establish political collaboration between the Communist Party and the Christian Democracy Party, aiming to bridge ideological divides and promote national unity during a time of political instability.

17. Martinelli, *Vajont*.

18. In the process of erecting the new memorial, the old tombstones were lost, perhaps even destroyed. This story is masterfully uncovered by Lucia Vastano in her book *Vajont, l'onda lunga*.

19. Chiara Baldi, "La proposta del ministro D'Incà: un monumento per ricordare Tina Merlin," *La Stampa*, November 5, 2021, https://www.lastampa.it/cronaca/2021/11/05/news/la_proposta_del_ministro_d_inca_un_monumento_per_ricordare_tina_merlin-385815/.

20. Giacomo Braulin, "Le farfalle del Vajont. La memoria a sessant'anni dal disastro" (master's thesis, University of Roma Tre, 2022–2023).

21. Citizen humanities are part of the broader field of citizen science, aiming to involve the public, particularly nonacademics, in the production of knowledge. A specific field of citizen humanities is public environmental humanities. For more on this, see Thom van Doren et al., "Developing the Public Environmental Humanities: Challenges, Opportunities, and Lessons," *Resistance. A Journal of Radical Environmental Humanities* 11, no. 2–3 (2025): 6–44.

CONCLUSION

1. White, *The Organic Machine*.

2. Donald Worster, *Rivers of Empire: Water, Aridity, and the Growth of the American West* (New York: Pantheon Books, 1985).

3. Scott, *Seeing Like a State*.

4. It is impossible to account for the intellectual debt to so many scholars. I mention, then, only the works that have most deeply influenced my work: Joan Martinez Alier, *The Environmentalism of the Poor: A Study of Ecological Conflicts and Valuation* (New Delhi: Oxford University Press, 2005); Laura Pulido, *Black, Brown, Yellow, and Left: Radical Activism in Southern California* (Berkeley: University of California Press, 2006); David N. Pellow, *What Is Critical Environmental Justice?* (Cambridge: Polity Press, 2018); Robert Bullard, *Dumping in Dixie: Race, Class and Environmental Quality* (Boulder, CO: Westview, 1990); and Julie Sze, *Noxious New York: The Racial Politics of Urban Health and Environmental Justice* (Cambridge, MA: MIT Press, 2007).

5. Rob Nixon, *Slow Violence and the Environmentalism of the Poor* (Cambridge, MA: Harvard University Press, 2011); Serenella Iovino, *Ecocriticism and Italy: Ecology, Resistance, and Liberation* (London: Bloomsbury, 2016); and Stefania Barca, *Forces of Reproduction: Notes for a Counter-Hegemonic Anthropocene* (Cambridge: Cambridge University Press, 2020).

6. Barca, Forces of Reproduction.

INDEX

Publisher contact:
The MIT Press
Massachusetts Institute of Technology
77 Massachusetts Avenue, Cambridge, MA 02139
mitpress.mit.edu

EU Authorised Representative:
Easy Access System Europe, Mustamäe tee 50,
10621 Tallinn, Estonia
gpsr.requests@easproject.com

Printed by Integrated Books International,
United States of America